# Echoes of Our Origins

# Echoes of Our Origins

## Baboons, Humans, and Nature

Shirley C. Strum

with Cassandra Phillips

JOHNS HOPKINS UNIVERSITY PRESS

Baltimore

 Published 2025
Printed in the United States of America on acid-free paper
9 8 7 6 5 4 3 2 1

Johns Hopkins University Press
2715 North Charles Street
Baltimore, Maryland 21218
www.press.jhu.edu

Library of Congress Cataloging-in-Publication Data is available.

A catalog record for this book is available from the British Library.

ISBN 978-1-4214-5203-6 (hardcover)
ISBN 978-1-4214-5204-3 (ebook)

A catalog record for this book is available from the British Library.

*Special discounts are available for bulk purchases of this book. For more information, please contact Special Sales at specialsales@jh.edu.*

EU GPSR Authorized Representative
LOGOS EUROPE, 9 rue Nicolas Poussin, 17000, La Rochelle, France
E-mail: Contact@logoseurope.eu

# Contents

# Preface

This is a book about baboons. Don't stop reading. I can't promise you will end up loving them as I do or be inspired by them as I have been, but you owe it to yourself to keep an open mind. It is also a book about me and the ways in which watching baboons has made me who I am. I wanted this book to be about baboons, not me, but I've come to know that, even in science, observations are made by human eyes and interpreted by human minds. That means you need to know me, at least a little, to make sense of why I think baboons are the way they are. And I hope by understanding who I was, and am, and how I changed, you'll get to know baboons almost as well as I did. This is not a memoir, per se, but I am on each page.

Fifty years is a long time. I never imagined that baboons would become my life's journey. In fact, after writing *Almost Human* I thought I didn't need to write another book. But I continued with my study of baboons after moving them to a safer but less hospitable habitat. This was an amazing opportunity to conduct an experiment with no precedent to guide me. I already knew the animals well. Now I would track their every move as they adapted in real time to a very different place. I watched as they confronted new challenges, and I learned how they coped and survived. As well as I thought I knew them after 13 years, they kept surprising me with new, unexpected behaviors. My ideas were forever needing revision, not only about baboons but about nature and evolution, and how science seemed to have hemmed itself in with its rigid methods.

If I hoped to truly understand baboons, I needed to go back to basics and work my way up from scratch. What *are* baboons? What can they teach us about other animals and about ourselves? I was trying to know the baboons' world using science, but the tools I had constrained me; often I was left with more questions than answers. What *is* science in the wild? Was there a better way to do science? And how was I supposed to align my findings with evolutionary theory? I'd begun to realize that baboon behavior didn't always square with what I'd learned about evolution. Is evolution actually about survival of the fittest? I'd seen baboons make mistakes that cost them dearly, yet they survived and had more babies. I thought of the demarcation line created by our culture—people living on one side, wild animals on the other. Yet today humans dominate both sides of that line, so where do we all fit in? Were people ever outside nature? And where, in relation to nature, are they now? Clearly, what I'd seen and thought about needed sorting out. This book is meant to do that.

I have struggled with how to present my argument, not just my interpretations of the things I saw but which words to use to describe them. In the end, I decided to unfold my discoveries as I made them, one step at a time. You will see baboon life through the animals I came to know and through my eyes as my understanding of baboons grew. I revisit early experiences—described in *Almost Human*—when my thoughts first took a heretical turn, but I present them now with a perspective sharpened by hindsight, experience, and study. The end of the journey informs all that came before, and an entirely new story takes shape.

But what is "the story?" I use many human words to portray baboons, necessarily and with good reasons that I explain. The word likeliest to daunt the reader is "complexity," but I use this term because I've yet to find another that better captures what I mean.

My husband tells me (and my editor concurs) that complexity will send everyone running for their life rafts. I hope not. I promise a slow immersion into the baboon ways that led me to complexity. I suspect that, in the end, you'll agree that it's a perfect, even lucid, descriptor for the baboon gestalt. Why does it matter? Well, you might wonder why baboons, smart as they are, are still baboons, and why humans are so different. Complexity may be the answer. Baboons communicate with "language" that's acted out, not spoken, and their social rules are sometimes onerous. Collaborative effort goes into the meeting of daily needs, but not always smoothly. It's all intertwined, and that makes it complex. Between their social and survival agendas, there's not enough time in a baboon's day to write books or build homes. The fact is, baboons have gotten along extraordinarily well without language or tools. But don't worry. I'll be there, holding your hand the whole way.

I want to remind you that these are my baboon observations, my interpretations, and my speculations. Some may be controversial. It remains to be seen whether future generations of scientists will prove my recent interpretations correct. Others, belatedly, have come to agree with many of my earlier ideas. Now a shift is underway in biology, genetics, anthropology, philosophy, and history of science, toward a rethinking of evolution.[1] I believe my baboon case study aligns with this ferment, and as such is relevant, timely, and quite possibly useful.

With that, I hope you find some fascination in the secrets that baboons have shared with me.

Shirley Strum
Nairobi, Kenya
November 5, 2024

# Color Plates

11. The translocation was a risky experiment, but it worked and was a model for how to do it right. (Dr. Shirley C. Strum.)
12. A large male baboon at sunrise on the top of the sleeping rocks. (Dr. Shirley C. Strum.)
13. The later team of Kenyans, doing inter-observer reliability tests. (Dr. Shirley C. Strum.)
14. The baboons prefer fresh green grass shoots over more mature grasses. (Dr. Shirley C. Strum.)
15. An infant eats the fruit of *Opuntia stricta*, the cactus that invaded the area. (Dr. Shirley C. Strum.)
16. The troop uses a game trail from the sleeping rocks; we can follow if we lose them by tracking their footprints in the soil. (Dr. Shirley C. Strum.)
17. When they were younger, Carissa and Guy often came to watch baboons with me. (Dr. David [Jonah] Western).
18. Trees are a good place to play for a change. (Dr. Shirley C. Strum.)
19. Jonah watching baboons with me at Namu's favorite sleeping site. (Dr. Shirley C. Strum.)
20. What if baboons wore hats—the uses for material culture. (Dr. Shirley C. Strum.)
21. Twala women enjoy dancing, whether for themselves or for visitors. (Dr. Shirley C. Strum.)
22. A male baboon jumps between sleeping rocks early in the morning at White Rocks. (Dr. Shirley C. Strum.)

# Echoes of Our Origins

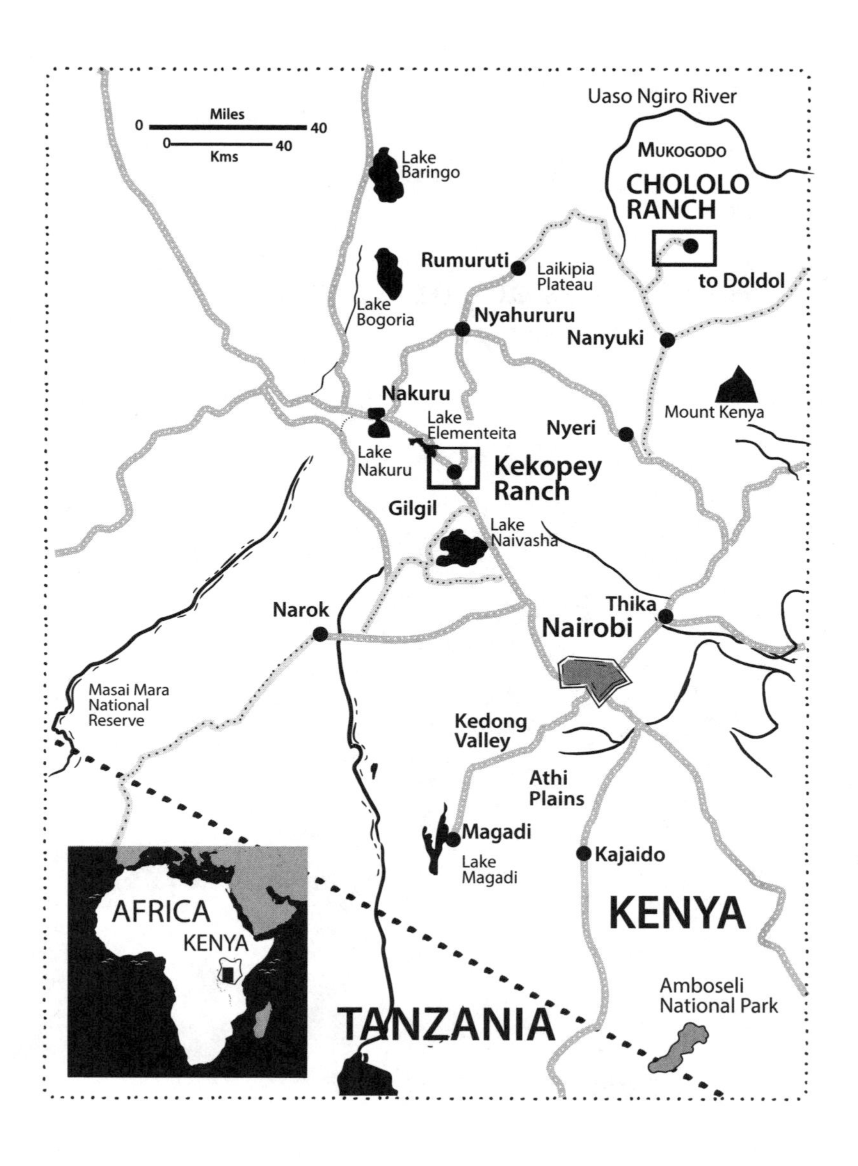

Miles
0
40
0
40
Kms
Uaso Ngiro River
Mukogodo
CHOLOLO
RANCH
to Doldol
Lake
Baringo
Rumuruti
Laikipia
Plateau
Lake
Bogoria
Nyahururu
Nanyuki
Mount Kenya
Nakuru
Lake
Elementeita
Nyeri
Lake
Nakuru
Kekopey
Ranch
Gilgil
Lake
Naivasha
Thika
Narok
Nairobi
Masai Mara
National
Reserve
Kedong
Valley
Athi
Plains
Magadi
Lake
Magadi
Kajaido
AFRICA
KENYA
KENYA
Amboseli
National Park
TANZANIA

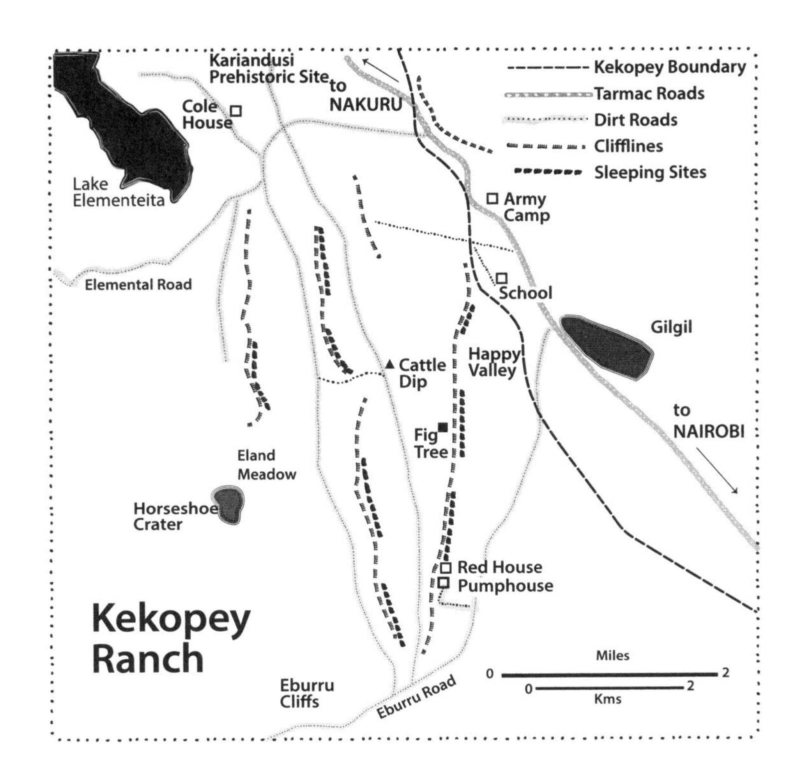
Kariandusi
Prehistoric Site
to
NAKURU
Kekopey Boundary
Tarmac Roads
Dirt Roads
Clifflines
Sleeping Sites
Cole
House
Lake
Elementeita
Army
Camp
Elemental Road
School
Gilgil
Cattle
Dip
Happy
Valley
to
NAIROBI
Fig
Tree
Eland
Meadow
Horseshoe
Crater
Red House
Pumphouse
Kekopey
Ranch
Miles
0
2
0
2
Kms
Eburru
Cliffs
Eburru Road

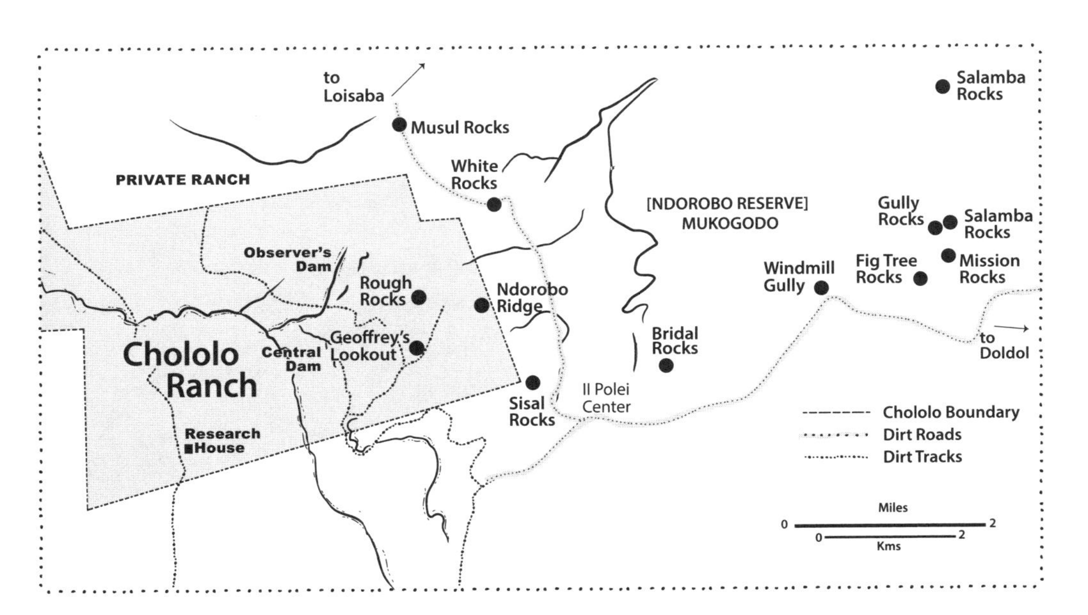
to
Loisaba
Salamba
Rocks
Musul Rocks
White
Rocks
PRIVATE RANCH
[NDOROBO RESERVE]
MUKOGODO
Gully
Rocks
Salamba
Rocks
Observer's
Dam
Rough
Rocks
Ndorobo
Ridge
Windmill
Gully
Fig Tree
Rocks
Mission
Rocks
Chololo
Ranch
Central
Dam
Geoffrey's
Lookout
Bridal
Rocks
to
Doldol
Sisal
Rocks
Il Polei
Center
Chololo Boundary
Dirt Roads
Dirt Tracks
Research
House
Miles
0
2
0
2
Kms

PART I

# Decisions and Revisions

## Chapter 1

# Why Baboons? (1972)

KENYA 1972: It's almost a cliché, people saying that Africa changes you forever. But I was heading to Africa to do fieldwork for my PhD; personal transformation was not on my agenda. Then, on my first day in Kenya, I found myself standing on a windswept precipice overlooking the Great Rift Valley. I did feel something beyond the expected wonderment, a sort of gasp of the soul. I ached to be down there, and yes, it did seem something inside me had shifted. Still, if someone had told me then that this place was where I'd spend most of my life, I would have laughed . . . or maybe cried. Nor did I imagine falling in love with baboons, or that baboons would have me questioning so much of what I'd learned.

I had landed in Nairobi after two days of hard travel and was headed for a remote research outpost. There, I'd study a "troop" of olive baboons and learn what they might tell us about the origins of human behavior. I would fulfill the fieldwork requirement for my PhD in Anthropology at UC Berkeley and be on my way to an academic career. I had done my due diligence: scoured the scientific literature, observed captive monkeys, delved deeply into all things Kenya, and assured my worried parents I'd likely survive the experience. Though it may seem improbable today, baboons were esteemed study subjects back then, the choice for models of early human evolution and yet to be viewed as obnoxious pests.*

---

My ride from the airport is with a fellow grad student, soon to leave the field and relieved to be going—a small red flag. He picks me up in an old VW van and we head north out of Nairobi. As outlying settlements thin to open range, the road gradually ascends to the top of the escarpment, a sheer cliff face that drops precipitously to the floor of the Great Rift Valley. Though excited and full of questions, I'm struggling to keep my eyes open. The van pulls into a clifftop turnout, and suddenly I've never been quite so awake.

Nothing has prepared me for this. Before me, a breathtaking green-gold expanse spreads clear across the curving horizon, dotted with broad acacia trees and veined with rocky streambeds, gently contoured like a quilt thrown over a bed and not yet smoothed. It all seems to shimmer with life. Two extinct volcanic craters loom in the distance, though I am yet to know their names.† On a clear day, I'm told you can see from this spot all the way to Nakuru, more than 75 miles away. I'd I had read somewhere that deep down we retain a sort of cellular memory of the savanna, humanity's birthplace, where we all began. I think I can feel stirrings of this memory.

Below us, the two-lane road zigzags to the bottom. I doubt this road could accommodate a van and bus meeting from opposite directions. Even worse, we're driving English style, on the "wrong" side of the road, skirting this sheer precipice as we make our way

*Their reputation has suffered in recent decades, thanks to media focus on baboon-human conflict.
†Longonot and Suswa.

to the valley floor. I hold my breath but can't take my eyes off the sight before me.

We reach the valley floor unscathed, and I get my first close-up look at the real African savanna. I have to remind myself that what I see wasn't what was here millions of years ago, before scattered settlements and rutted roads. Up top, I barely saw these imprints of humanity. We come upon a herd of giraffes, a sight that puts the zoo experience to shame. Here in the wild, up close, giraffes are implausibly massive, yet so benign—gentle giants with fantastically long legs and necks. I lean out the car window to get a better view of their heads. Nothing in their features suggests danger. Like Disney characters, their long lashes frame big, friendly eyes. Several adorable babies peer at me from the safety of their mothers' sides. Variations in their horns let me quickly tell them apart, a good sign. Maybe I'll be able to identify individual baboons.

We pass zebras and impalas on both sides of the road. Now I understand how zebra stripes can function as camouflage; the animals vanish as we leave them behind, blending into the wavy heat haze. We drive on through more wild animals, including tiny dik-diks, a diminutive antelope with a funny mobile nose. I see several pairs as we pass, and I know from my Kenya crash course that they mate for life. I'm glad to be in the car as we pass near the larger of these wild animals, the giraffes and zebras, even though I know they won't hurt me. But I am green, and this is the first time I've experienced wildlife living free. Will I have the courage to walk among baboons, exposed and vulnerable, in a place known for its wild animals—including predators? This is the very thing that unnerved my colleague, who's already warned me about "killer warthogs."

Finally, we turn off the main road onto a dirt track that leads to the "research house," my new home on a 45,000-acre ranch called Kekopey. Why, I wonder, are there tall poles with lines strung between them? I've been told the house has no power. Then comes a jolt. A man is hanging, quite dead, from one of the lines.* We drive on as if this is an everyday occurrence, but my breath catches in my throat. I try to make sense of what I've seen in the last several hours, on one hand the ethereal beauty of the savanna and its wondrous wildlife, on the other, an unsettling image of human horror. Over the next five decades, spent mostly in Africa, I will continue to find the natural world more hospitable, and kinder, than the world occupied by my own species.

I am keen, the next day, to get into the field and start my study, despite a nasty case of jetlag. It just happens to be my twenty-fifth birthday. We take the VW van to the site and crouch inside, peering at a cluster of baboons through binoculars. It's my first encounter with this troop, as baboon groups are called, waggishly dubbed the Pumphouse Gang.† I try to make sense of the 59 animals, maybe 200 feet distant. Using binoculars this first day, I can only distinguish big from little baboons, not much of a feat since adult males weigh about 50 pounds and infants just two pounds. Where are the adult females? Adult male baboons are almost twice the size of adult females, but immature males are the size of females. What's different about them, I wonder.

I expected more action, tussling, maybe a dash of conflict. Instead, I see mostly quiet resting with some feeding. The closest knot of baboons is eating grass. I make out a single baboon giving a series of soft low grunts. Others reply with their own grunts. What does this mean? Baboons don't have "language" as we know it, but they communicate with sounds, postures, gestures, and facial expressions. I found the "grunt round" comforting. Did the baboons?

*When I ask later, I'm told that his death was a probable suicide. No less discomfiting for that.

†A nod to Tom Wolfe's "new journalism" book, *The Pumphouse Gang*, about renegade Southern California surfers, and fitting because a small pumping station lies near a favorite baboon sleeping site.

Other social interactions defy interpretation. A large male baboon approaches another male. He takes hold of the other's hips while rapidly smacking his lips. The approached male extends a hand—perhaps a friendly gesture—and glances up at the fellow who grasped him. Then it is over; the greeter walks away. Is this dominance affirmed? If so, why did he leave instead of "displacing" the other male by forcing him to give up his spot, a behavior I had read about? Were encounters between males always so frictionless? Not at all, I was soon to learn.

I spot a mother cuddling her very young jet-black baby. Baboon babies are born a color that contrasts with adults. I'd learned that infants' special color might protect them from aggression, or possibly make them attractive "super stimulants" to adult baboons. Perhaps it does both. In any case, I see proof of earlier observations about the monkey fixation on infants. One baboon, then another, and another, approach the mother and baby. Mother rebuffs most and permits just a few to touch her infant, *if* they first groom her. Grooming is common across all primates, a probable outgrowth of hygienic practices that now serves as a multifaceted social tool. It's fascinating to see the groomer's intensity and care, sifting hair like ours, not fur like nonprimate mammals, going row by row, picking off parasites and debris. But only certain baboons are allowed close enough to groom. What else is going on?

---

I had prepared for this day by reading every published study about primates, baboons, and animal behavior. Back then it was possible because the scientific literature was skimpy—a situation soon to change. I had immersed myself in ethology, the evolutionary approach to animal behavior. I had spent hours observing gibbons, the smallest ape, as well as langur monkeys, housed in large outdoor enclosures in the hills above the UC Berkeley campus. By the time I graduated to patas monkey, I'd invented my own shorthand for quickly recording behaviors. This would come in handy with the baboons.

But why me, and why baboons? I fit into the times of Jane Goodall and Dian Fossey, both picked by Louis Leakey, the legendary paleoanthropologist, to study chimps and gorillas because their minds weren't cluttered by "science." Yet I was unlike them: a confirmed city girl, a trained scientist, and not a social loner. I'd been shaped by my time at UC Berkeley, during the thick of the Vietnam War and the militant early days of second-wave feminism. Both causes captured my attention and found me marching in the street on more than a few occasions.

I identified more as rebel than hippie back then, but we all shared a disgust with the violence we saw around us, in the killing fields of Southeast Asia and also, too often, against women by men. Why did violence seem such an intractable part of being human? I wanted to understand this violent impulse. Why had it persisted despite our vaunted evolutionary advantages and our laws and spiritual teachings that insist we should do each other no harm?

My choice to study the evolutionary persistence of violent behavior was personal as well. I am an only child of Holocaust survivors, and my childhood bore scars of human aggression. Unlike other "survivors" we knew, my parents never spoke of their experiences. I grew up knowing they'd endured something unspeakable and that many relatives were lost, but little more. I finally learned from others that my father had been at Auschwitz, my mother at Bergen-Belsen. That was it. In the absence of details, my child's imagination filled in the blanks. We often speak of "triggers" these days. When my elementary school screened a film about World War II, you could say I was triggered by a sequence showing people in concentration camps. I ran from the room, screaming. Thereafter, I was excused from war films. After my father died, my mother was finally willing to talk about the camps, but she was in her eighties and afflicted with dementia. She was convinced she'd been on

Schindler's list. I knew this to be untrue, but I let her believe it was so. Her camp experiences were lost to me forever. But their trauma left its mark. Freight trains at railroad crossings still give me a sharp surge of panic.

And so it was that I sought the best scientific approach to the study of human behavior, a path that might lead me to the origins of humanity's violent streak.

To say my parents were anxious would be an understatement, and for this only child that meant constraints. When the time came for me to go to college, I wanted to get as far from our San Diego home as my parents would allow. Which is not to say that UC Berkeley was a compromise. Berkeley was an exciting place in the mid-1960s, a hotbed of intellectual ferment as well as protest. But even at Berkeley, individual dorms weren't yet coed. I would slip away from my women's dorm at midnight to wander down Telegraph Avenue, taking in the scene and catching snippets of heated conversations in cafés and bookstores that never seemed to close.

I sampled classes in biology, sociology, psychology, and even criminology, but none seemed to address those deepest roots of humanity's problem with violence. Then I took an introductory physical anthropology course from Sherwood Washburn, a star in his field, now known as the "father of American primatology."* He was open and approachable, a short man with a thick shock of gray hair and horn-rimmed glasses that had way of slipping down his nose.

In that first class you could hear a pin drop in the cavernous lecture hall as Washburn elucidated human evolution to more than 1,000 students. He made abstract ideas feel like living things and

*Washburn, called Sherry by his friends, is more famously known as "the father of modern physical anthropology" (Sherwood L. Washburn, "The New Physical Anthropology," *Transactions of the New York Academy of Sciences, Series II*, 13, no. 7 [1951]: 298–304), now called biological anthropology.

nimbly synthesized seemingly disparate concepts. Genetics had just revolutionized our ideas about evolution. Washburn suggested that an anatomist would need to see a creature using its body in the natural world to understand the evolution-driven functions of muscles and bones. This was a highly unorthodox idea, and I liked his fearless interrogation of the status quo. He disdained "scientific" speculation about early human evolution and urged a greater commitment to facts—and to their discovery. In a memorable turn of phrase, he said, "Behavior doesn't fossilize." He was convinced that the study of our close evolutionary cousins, nonhuman primates in the wild, could unlock the origins of human behavior and replace speculation with evidence.

It was always a bit unsettling to talk to Washburn in his office because many topics that excited me were "dead issues" to him. In truth, I mostly listened and avoided contradicting him. He would quickly shut down any student who tried to question his ideas, but it didn't smart because he did it gently. I learned from watching him that "silverbacks," at least those of a certain stature, may decide they have *all* the answers. I realized how great a man he was, scientifically, when later I was asked to edit a volume in his honor and read tributes to him from around the world.

I had a major and a mentor. Well, almost. Anthropology is a multifaceted discipline, and certain of those facets did not appeal. I couldn't feature myself spending years, months, or even days wielding a toothbrush in a sunbaked riverbed, digging up fossils or pottery. Nor did I like the idea of embedding myself with a group of people and observing them as if they were insects in a bottle. But here was the Washburn option, the study of primate behavior. I was lucky to land at one of two universities that offered serious primate study programs, the other being Harvard, led by one of Washburn's former students. I decided to stay at Berkeley and pursue a PhD.

For my thesis work, I proposed a study of patas monkeys in Kenya, focusing on their unique social organization. Like baboons, patas monkeys (*Erythrocebus patas*) live away from the safety of forest trees, making their society a different savanna test case. They'd been the subject of only two short field studies, and more could be learned.[1]

But Washburn wouldn't have it. "Patas have been done," he said. Instead, he wanted me to test a widely accepted "model" of baboon behavior that he'd helped devise. Baboons made perfect study subjects, he said, because their society looked to be structured by male dominance and aggression—the basis of the baboon model—much, alas, like ours.[2] Since human behavior seemed to echo the ways of these primates, new insights were possible. I had little choice in the matter and trusted Washburn to know best. I would go to Kenya, watch baboons for 16 months, write up the results, get the PhD, then embark on a university teaching career. Little did I know . . .

---

Given how much we've learned about primates—the biological order that includes humans as well as monkeys, apes, and prosimians—it's surprising how little was known about them in the early 1970s, especially species living in the wild. The evolutionary perspective on animal *behavior* was particularly modest and rudimentary. This began to change when Washburn launched a new wave of primate studies in the 1950s and 1960s.[3] This was when Louis Leakey, then at the peak of his fossil hunting fame, sent young, untrained Jane Goodall to Tanzania's Gombe Stream Reserve. Her groundbreaking study of chimpanzees would show striking behavioral similarities between chimps and humans.[4]

Chimps would eventually replace baboons as the model for the "lost" behavior of our earliest human ancestors. But early in my graduate career, baboons seemed likelier analogs of early human evolution. The reasoning was that, like us, they'd left the forest—still the primary primate habitat for apes and monkeys alike—and ventured onto the savanna. Out in the open, exposed to an array of new predators and well-adapted grassland competitors, a puny primate species would need a new set of strategies for survival. In fact, few primate species actually managed to survive on the savanna. Baboons and humans were the most successful. Even recently, baboons were thought to outnumber humans on the African continent—as ape populations continue to dwindle.

By the time Washburn assigned me to baboons, I considered them well studied compared to other wild primates. Work done by Washburn, Irven DeVore, and Hall had made baboons the most closely examined wild primate species at the time. Between 1959 and 1961, Washburn watched yellow baboons (*Papio cynocephalus*) in Kenya's Amboseli ecosystem and DeVore, a Washburn protégé, researched olive baboons (*Papio anubis*), the type I would be tracking, in Nairobi National Park.[5] Hall, an English ethologist/psychologist, studied chacma baboons (*Papio ursinus*) in southern Africa.[6]

The so-called baboon model emerged when these three scientists compared notes from their individual studies. The similarities were striking, especially since the baboons they observed were different species living in different parts of Africa.[7] All of these baboons, no matter the species or habitat, formed cohesive societies and spent their days feeding, socializing, and resting as a group. Each troop slept together in trees or on rocks to thwart nighttime predators. The groups were large by primate standards but rarely numbered more than 80. Mature females outnumbered mature males, while "youngsters" made up the majority of the group. At sexual maturity, most young males left and joined a new troop while females remained in the group where they were born, termed their *natal* troop.* Once migratory, males rarely stayed for more than five years in the same troop.

Something else excited Washburn. Drawing on separate datasets about baboon anatomy and behavior, Washburn concluded that

*Thus avoiding or at least minimizing inbreeding, though it's doubtful they do so intentionally.

form (anatomy) and function (behavior) had coevolved for life on the savanna. The reasoning went like this: A male is nearly twice the size of the average female, and unlike the female, has imposing canine teeth that get sharpened each time he opens and closes his mouth. Males also sport a thick mantle of hair around their head and shoulders that bristles during aggression, giving them bulk and a look of ferocity. The conclusion: Evolution equipped male baboons with the *anatomy of aggression*.

But why? Surely this fearsome anatomy would help deter savanna predators. In the forest, a primate group could be safe in the trees. On the open savanna, safety and defense posed far greater challenges. These early baboon watchers had observed males advance as a phalanx toward lions, hair erect, teeth bared, white eyelids flashing, barking fiercely. Even lions went looking for easier prey.*

The group structure also helped ensure survival. Not only was there safety in numbers, but also the advantages of a social order revolving around the large males. Each adult male had a rank in a dominance hierarchy and a role to play. As in a military troop, hierarchy was a hedge against chaos, which can cost lives.

The model offered another piece to this evolutionary story. Targets of bluff and aggression could include males *within* the group. Scientists had observed male baboons menacing other males when competing for food, receptive females, and dominance. They saw a stable male dominance hierarchy as a tool to manage aggression. That meant fewer daily fights over resources and rank since each male knew his place.

When males competed for limited resources, the lower-ranking male deferred to the higher-ranking one in a behavior called

*To see what happens when it is just one baboon against a lion, see shashaenright, "Lion vs. Brave Baboon Fight—West Serengeti Safari," YouTube, September 13, 2009, 3 min., 4 sec., https://www.youtube.com/watch?v=4ebd36p4zkw.

displacement. However, alliances of several lower-ranking males could gain advantage over one who ranked higher without changing any male's individual rank. Thus, said the model, the male hierarchy fostered a stable and predictable structure for the group. Males also policed the group's internal disputes. The troop, and the species, benefited, since the most capable males fought their way to dominance then monopolized resources, including sexually receptive females. The best genes from the "fittest" males made their way into the next generation—or so it seemed—and it made for a compelling model.

A special formation seen during group movements bolstered the baboon model and seemed to support its evolutionary basis. It happened when the troop moved en masse across the open savanna. Dominant males encircled mothers with small infants at the center of the troop, acting as a protective shield to the most vulnerable and important group members. Other females and immatures formed the outer rings of the formation, with the lowest-ranking males at the perimeter. If a lion made a surprise attack, these peripheral males would be fodder. However, when a predator approached, dominant males moved to the front to challenge the would-be attacker.

I could understand Washburn's excitement. The baboon model told a compelling evolutionary story at a time when there was little evidence suggesting how behavior figured into natural selection or how behavior and anatomy might be linked. But when it came to female baboons, the model got fuzzy. Females were viewed as nonaggressive, nonhierarchical, and lacking structure in their social connections. Their role was to nurture the next generation, a vital and time-consuming job. A female's rank seemed to wax and wane with her sexual cycle. In the model, she briefly attained high status if she mated with a dominant male but lost her standing once he moved on.

Both scientists and the public—because this research got attention—embraced the baboon model; it made so much evolutionary (and, yes, cultural) sense. Yet there were dissenters. Thelma Rowell, watching forest olive baboons in Uganda, saw males be the first to flee to the treetops when they sensed danger.[8] Vulnerable members of the troop, including mothers and infants, were left to fend for themselves. Tim Ransom, studying olive baboons at Goodall's site, never saw the classic troop movement with dominant males near the center.[9] By the early 1970s, when I was in grad school, a few newer studies questioned aspects of the baboon model.[10] Getting it right mattered, because the model had traction. It was serving as a template for studies of other nonhuman primates, and, more importantly, for scenarios of human evolution. Though it was not discussed at the time, I now wonder if the model might also have offered an evolutionary rationale for unequal sex roles, then the norm in Western society.[11]

Admittedly, Washburn switching me from patas monkeys to baboons came as a shock, though I got the value of putting the baboon model to the test. But if I was stuck with baboons, I wanted to test the model with all the rigor I could muster. I'd borrow new data collection methods used to study animals in captivity and apply them to baboons in the wild. My hope was that these more objective, almost clinical, methods would control for possible biases embedded in the earlier studies. I resolved to identify and follow each troop member, even the shyest, most retiring female. It never made sense to me that only large males got identified, leaving the rest of the troop a sea of nondescript faces. Surely a more granular picture of baboon society would yield something new and worth knowing.

I am reminded of something Washburn often said: "Today's science is tomorrow's superstition, because facts are only the best fit between current methods and reality." I took that message to heart.

Besides, at the time, neither Washburn nor I expected my findings to dismantle the image of baboons he helped create.

---

Many discoveries, adventures, and challenges awaited me. My first study led to new ones and a career with myriad twists and turns. And then, incredibly, 50 years had gone by. What happened was, I fell in love with baboons, as odd as it may seem given their, shall we say, spicy reputation, even then. But there was always something new to learn. Fresh insights built on each other, only to be met, quite often, with a bracing dose of resistance from peers. Mainstream science at the time had fixed ideas about baboons, evolution, the practice of science, and the question of animal mind. Many studies did not comport with my findings. My struggle as a woman scientist was not so much about being heard as being believed, especially as I began to see baboons using *alternatives* to aggression and rarely hurting each other, even when it looked like they might.

My story has a few personal twists as well. Baboons weren't the only ones I fell in love with, nor was motherhood on my agenda until it was almost too late. Then a medical crisis nearly ended it all. You will hear of baboons I rescued from near-certain death, and about moving them to a distant habitat where their survival was equally uncertain. Conservation became an abiding concern and led to deep involvement with Kikuyu farmers who settled on Kekopey, and later with Maasai herders who shared the baboons' new home. With each set of people, we have searched for ways they and the baboons can coexist and even support each other. And then there's Cape Town. I was called in to help with a baboon invasion that threatened to cause civil war between pro- and anti-baboon factions. There will, however, be no killer warthogs, though lions and leopards make occasional unwelcome appearances.

Mine is not a simple story to tell because it spans half a century that brought changes to science as well as to my interpretations of baboons. Captivated by things I saw that didn't fit predictions, I followed my nose rather than the scientific culture. Yet even today, uncertainties linger and always will, unless baboons learn to talk, or we learn to read their minds. Still, the baboons taught me patience because, over time, if I watched very carefully, they would often "tell" me the *why* of their baboon ways. Fifty years is a long time and includes the Human Age, the Anthropocene. The baboons slowly gave up their secrets because I didn't give up, in spite of all that stood in my way.

But I get ahead of myself. First, I needed to get to know the baboons of the Pumphouse Gang.

Chapter 2

# Learning Baboon (1972–1973)

**I LOVED TRAMPING** out in the early morning, over the dusty dirt track that led from the rustic research house to the baboon's sleeping rocks when they slept nearby, which was often. I'd been considered such a "city girl" in my university milieu that several friends took me aside and wondered if my doing fieldwork in Kenya was such a good idea. But surrounded by baboons and other wildlife, and almost without people, I thrived. In the open spaces and big skies of Kekopey, insulated from Kenyan politics and white racism, my romance with nature began. Kekopey was a working cattle ranch, but home to more wildlife than cattle. The owners, good people, provided lodging for researchers in a remote former manager's house called Kiserigwa. Built in a U shape, it had two bedrooms and a kitchen with a wood-burning stove that also heated water for the bathroom and kitchen. I shared the cottage with a stream of fellow researchers over the better part of a decade. Sporadic home improvement efforts resulted in real end tables replacing cardboard boxes and threadbare seat cushions traded for new ones. The house was near a fault line cliff that bisected the ranch; beneath this ledge ran the dirt track that also served as a road when the baboons slept farther away and I needed to drive to where they were. Another fault scarp ran across the ranch to the west of this one. The baboons sometimes slept there, too.

It felt like I'd landed in a dreamscape. From the baboons' main sleeping site, I could catch a glimpse of distant Lake Elementeita, an alkaline "soda" lake often covered with pink flamingos. To the west was a volcanic cinder cone named Lord Delamere's Nose because it resembled an Englishman's face looking skyward. I could see up and down the Great Rift Valley, a stunning panorama that included nearby volcanic badlands, home to buffalo and black-and-white colobus monkeys who lived in the rubble's cedar forest. A male colobus monkey occasionally appeared in a fig tree near the baboons' favorite sleeping site, quite an accomplishment for an arboreal monkey who would need to traverse a large open space to get there. He'd arrive just as the figs ripened, gorge himself, then disappear as mysteriously as he had come.

Grasslands dominated the scene. Kenyan savannas are highly productive because of their double rainfall seasons brought by Indian Ocean monsoons moving first north, then south in this narrow zone of Africa. The "single" rainfall season in the rest of the continent produces tall, rank grasses that persist during a single long dry season. That's a challenging time for everyone, including baboons not lucky enough to live in this unusually verdant swath of the savanna.

Wildlife was everywhere. More than three times the size of Manhattan, Kekopey was rich in grazing areas, thanks to water piped in from hot springs in the hills to troughs throughout the ranch. Large fenced "paddocks" divided the property into sections and controlled cattle movements and grazing. But wildlife paid no heed to fencing. Zebras found gaps in ravines and went under, eland (a species of hoofed mammal larger than a cow) jumped over, and impalas did both. Buffalo didn't venture far from the badlands, a good thing since they are dangerous when spooked, and males spook very easily. I found the variety of wildlife astonishing: two kinds of reedbuck, Thomson's gazelles, steenbok, duikers, aardvarks, Cape hare, springhare, and more. Elephants and giraffes once grazed here, but no longer, why I could only imagine. Leopards and lions had been trapped out to protect the cattle. I often saw jackals, bat-eared foxes, and an occasional hyena when I was walking with the baboons. At first I had felt wary even of the baboons, whose behaviors were so new to me. But my fears melted away as I got to know them. I also realized that local wildlife had better things to do than bother me.

Even the elements were wild. At Kekopey, a stunning drama of natural forces played out almost daily. In California, rain and fog were inconveniences. Here, alternating rains and heat dictated almost everything everyone did. Imagine watching baboons in a heavy downpour and trying to take notes on paper. My only defense was a basic rain poncho and leather safari boots with thin rubber soles that should have stayed drier than they did. I became intimately acquainted with mud, and I learned to read the signs of an impending downpour, just like the baboons. During storms, they huddled together under a tree or bush, backs turned against the onslaught. Babies kept dry and warm as they nestled in mothers' laps. I would have liked to join them.

---

Even though I'd done so much to prepare, those early days in the field had my head spinning. I saw baboons behave in ways I hadn't read about and began to sense that we'd only scratched the surface of their world. By the end of that first day, I felt completely upended—excited but a little lost, fascinated yet overwhelmed. Still, I'd made two crucial decisions before coming to Kenya: I had to get out of the van so I could follow individuals more easily, and

I had to identify each and every baboon in the troop. My Berkeley colleague, now handing his post off to me, seemed uncomfortable even observing baboons from the van.

While the animals had grown used to a parked van, they'd never experienced closeness to a human primate, on foot, right there in their midst. I had no idea how they'd react. Would they know I was female, and would it matter? I didn't have answers to so many questions—yet—but being a "woman anthropologist who studies primates" turned out to be important to both feminists and science.*

My two graduate student predecessors had erred on the side of caution, I thought. Close physical proximity to the baboons seemed so fundamental. I needed to know each animal and how it fit into the troop, and I had to recognize individual behaviors and learn what they meant. Only then could I begin to consider what baboons might tell us about human evolution. Staring at them from a van would not get my job done. Neither did I want to risk human-induced disruption of their behavior, or my own safety. A prudent approach was necessary. Soon after my colleague left and I was finally on my own, I drove to the baboons, parked, slowly slid open the van door and stepped out onto the savanna.

That day, I stood by the van and watched. The baboons went about their business. The next day I crept a few feet closer. I stopped, gauged reactions, and waited. Day by day, step by step, I made progress. Meanwhile, I observed and took notes. After weeks of gingerly approaching the baboons, I could sometimes stand quietly and watch from the periphery of the group. After several more weeks I had penetrated the outskirts of the group. Several months later, I was fully in, able to track an individual baboon through the middle of the troop, ever mindful of the reactions of nearby animals.

I very quickly learned that baboons have their own etiquette, expressed in behavior. This includes movements made toward or

*Later, the term "primatologist" came into vogue. I never warmed up to it.

away from another individual. An approach can ask a question, such as "May I groom you?" Backing away, called an "avoid," says, "Your higher rank takes precedence." A lipsmack sound and gesture says, "I mean no harm." The behavior or gesture must be made at an appropriate distance, in a correct manner and context.

I was beginning to pick up on these social cues. If I sensed I was making a baboon nervous, I stopped and turned away or even stepped away slowly, which should end the tension. Looking was an important part of any interaction. You can't interact with a baboon who refuses to look at you, and vice versa. It was an axiom that proved especially helpful when, later, a young subadult male tried to become dominant over me. When I turned my head away, he understood that I refused to engage and would not play his game. But he didn't give up; we made a funny sight as I walked in small circles, avoiding his attention, and he following in larger circles attempting to catch my eye. I also learned that I could use my clipboard as a sort of screen to block my gaze. This move came in handy if the baboons got edgy in situations where we found ourselves bunched together, in a thicket or at the sleeping site.

Generally, I maintained silence and kept myself apart at a respectful distance but this varied depending on the baboons' activities. I could get closest at the sleeping cliffs, where baboons would start their day and gather at day's end. But if I came within 15 feet of a foraging baboon, I'd cause my "subject" to sidle away. A rule I learned early on was always to maintain "baboon distance," meaning I kept as much space between them and me as they kept between each other in a given situation.

And so we established ground rules. The baboons, by every appearance, were acting like their usual baboon selves in my presence. "Habituating" the baboons taught me the valuable lesson of how to rein in my typically human self. I learned both patience and respect for the baboons that continue to this day. I'm bothered when a new graduate student speaks too loudly or crowds a baboon and fails to read its distress. If only *they* had been the ones to habituate these baboons . . .

---

As I got closer to them, I realized that I could use baboon size to distinguish males from females, adults from adolescents, and juveniles from infants. Baboons are primates like us; we share anatomical features and experience the physical world similarly. We both have forward-looking eyes, overlapping fields of vision, depth perception, color vision, and hands with mobile digits. We also share broad categories of behavior, among them mothering, play, nonverbal communication, and status awareness. I assumed that my perception of the baboon social world was not so different from theirs.

What I didn't yet know was how to use my primate vision to tell the baboons apart. Before switching to hamadryas baboons, my idol, Hans Kummer, managed to identify individual *Drosophila* flies (fruit flies) by dotting each with a different color of nail polish. Nail polish was not an option for me, though it might have helped.

As a person who can't tell horses apart, except by color, I had my work cut out for me. I began with the large males because they were fewer in number and, in some cases, strikingly distinctive. As I gained confidence, I used a process of elimination for the more difficult ones. That is, if it wasn't this one or that one, already identified, then it had to be this or that other one. I made terrible drawings of the males to help me, and they did, but they also reminded me of why I chose not to become a biologist. In an introductory biology class, we were asked to sketch basic animals. Mine rarely resembled an animal of any kind.

My identification techniques developed quickly. I learned that a distinctive baboon feature, the calloused bottom, told you an animal's sex. The hardened pads, technically called ischial callosities, are formed from hair, like your fingernails, but are much thicker. These callosities prevent pinching of the sciatic nerve so baboons can sleep sitting up without their legs going numb. Males and females have differently shaped callosities. Male callosities are a single swath under the tail. Female callosities are split down the middle, allowing sexual skin to swell when they're receptive to mating. It's a difference visible even in newborns and a handy way to tell prepubertal juveniles apart.

Rad was the first large male I could identify. I had come up with a system that broke down individual features and applied it to each male, and later to each female. These features included tail shape and angle; type and color of hair; shape of body, face, eyebrows, and nose; hands colored black or not; and scars or other singular markings. Rad had a distinctive tail that bounced like a metronome as he walked. But it was his nose that was unique, a Pinocchio-like attachment on his muzzle. He had very dark hair, and his eyebrows arched in a way that made him look, at least to me, like a "kind" baboon (I'd noticed that others with eyebrows that slanted up at the ends had a slightly evil look; see appendix 1). Rad was my Rosetta stone.

David was different. He was almost blond, resembling the female I suspected to be his mother. To me, he was beautiful. He was not quite adult and still unblemished; no scars ran across his muzzle, no patches of hair were missing due to bad condition or old age. Rad had long hair, but David was fluffy. His tail was short and curved, his brow was slanted neither down nor up, and his hands weren't black. Slowly but surely, the males revealed themselves to me, including the new immigrant, Ray, soon to be a central figure in my early understanding of the baboons.

Then I started on the females. Because there were so many, I looked first for individuals with special characteristics. I noticed that three adult females looked like sisters, sharing the same body type and with similar faces and tails. Only their hair color varied. They didn't seem very pretty, but maybe I wasn't the most qualified judge of baboon beauty. I called them the three ugly sisters, and I suspected that one was handsome David's mother.

Then there was Peggy, who was to teach me so much and later break my heart. Peggy's tail was straight and long, and it turned down at a ninety-degree angle from her back. No other animal had a tail at quite that angle, except her children. Her hair was dark with distinctive blond specks, and on her forehead was an unusual black V. One eye had an opaque milky look, perhaps due to an acacia thorn injury. Her partial blindness was evident when she held an infant aloft and toward her good eye.

Naomi and her infant Robin posed an instructive challenge. I had trouble identifying Naomi at first; she seemed almost generic. Then I noticed she was the only female always sitting by herself at the edge of the group. I realized it mattered where an animal physically situated itself in the group. Robin, Naomi's daughter, was already a transitional infant; her pink face was turning dark and splotches of brown peeked through black hair, particularly on the brow ridge.

Except for Peggy, eyes were not particularly helpful in identifying individuals, yet they are both captivating and important. Human eyes have white sclera with dark pupils, making it easy to tell the direction of a person's gaze. Most primates have dark sclera and dark pupils, making it hard, if not impossible, to know what they're looking at. Baboons are different. They have amber eyes with dark pupils. When a baboon looks at you, you know it. A direct stare is sometimes a threat, but not always. If I ventured a direct glance at a baboon, the eyes told me "someone was at home."

Infants were next. The youngest ones kept close to their mothers, which helped. As they matured, I could track changes in size, shape, and color. Finally, I tackled the juveniles, who all looked alike—brown, bigger of course than infants, with ears, hands, and penises, in the case of males, in better proportion to their bodies.*

*Male baboons are born with oversized penises that make them unmistakably male. As the animal grows, his member gradually becomes proportional.

In a matter of months, I was able to set aside my system. Each animal had become known to me as an individual. I recognized them as they probably recognized each other—by the "gestalt" of all these characteristics and, of course, by their faces, each of which I could now recognize as clearly as those of my friends.

When I started my field study, researchers did not routinely identify individual animals, except sometimes numerically. The idea was to avoid anthropomorphism at all costs. Nonetheless, Japanese primate scientists pioneered the practice in their early studies of Japanese macaques (1950s) and of chimpanzees in Tanzania (1960s).[1] But it was Jane Goodall who really captured the public's imagination by describing and naming individual chimps in her 1960s Gombe study.[2] Despite the sensation caused by her chimp family stories—or maybe because of it—most Western scientists continued to discount the value of knowing individual animals for another decade.

Due to the focus on large male baboons, Pumphouse males had been given names by the time I arrived. Few of the females or youngsters had. So I set about naming them all, not as an indulgence but as a practical necessity, the best way to keep track of the animals and the maze of behavior that unfolded each day. I used an alphabetic system. Offspring of a mother carried names with the same first letter. Thus, Peggy's children were named Portia, Patrick, and Pebbles. Variations occurred as the alphabet proved to have limitations and as generations matured. But at first it served me well.

---

At the end of each day, I could hardly wait to get back to the field the next day. At night, I dreamed about baboons. From the start, in those early uncharted days, I was smitten. So much about the baboons was taking me by surprise, and so many of the surprises felt like revelations. What most amazed me was their acceptance. Knowing I respected and adhered to their social rules, they relaxed and let me be among them. Their trust was a gift for which I was endlessly grateful. Without it, my study would fail. Watching the baboons met many of my needs—social, emotional, intellectual, and physical—and demanded little of me, except my undivided attention and polite compliance with their rules. Everything was new, everything was a puzzle to be solved. Finding any piece was thrilling. If this was *science*, I loved it.

I realized I had become a de facto member of the Pumphouse Gang during a chance encounter with another baboon troop, about six months later. I was following Pumphouse on their daily walk-about, taking notes as usual, when suddenly we came face-to-face with a wild and unfamiliar troop. I found myself within feet of these baboon strangers, and I had no idea what would come next. Would everyone scream and run? Would they hold their ground, maybe do a little mingling? Would there be fighting? Incredibly, we all kept our cool, even me. To the strangers, I was just another Pumphouse baboon; because Pumphouse accepted me, so did they. Now I could learn what happened when troops "collided." I watched, fascinated, as adolescent females got flirty with males in the other troop, juveniles played with each other, and adult males worked at keeping "their" females from "crossing" the line. After a few hours, we all went our separate ways. I can still feel the excitement of that morning.

There were, of course, episodes of aggression within Pump-house. I well remember the first one I witnessed. As I watched from my spot at the edge of the group, two unmatched males charged each other—David, nearly adult, and Carl, already aging. I have to admit that I found blond David gorgeous, while Carl could have been a preview of older David, with scars on his muzzle and less bounce to his tail. Both animals slapped the ground and threat-ened each other with eyelid flashes and loud pant grunts. So *this* was the male aggression I'd heard so much about. As the tension reached an almost unbearable pitch, David appeared the stronger contestant. What next? Carl suddenly took off! I assumed that was it, that Carl had lost his nerve and hightailed it away from the mighty David. But just as I jotted down "Carl, loser," he reappeared with a black infant clinging to his belly. He sat facing David, holding the baby like a shield. As if a switch had flipped, David backed away. Wow! What was that all about? Nothing in the literature described such an encounter.

I decided to start a list of "unusual events" like this, but I soon learned that events that had seemed anomalous really weren't. Infants often played a part in aggressive encounters between adult males, but in various ways. As these instances populated my notes, I remained mystified. It wasn't until later that I would fathom this behavior.

---

At first, my data gathering involved watching baboons and speaking my observations into a tape recorder. It was a thorough method, but it had a big downside: the need to transcribe. Each hour of speech required at least two of typing. Not sustainable. So I developed a new "check-sheet" (see appendix 2) that enabled the collection of "facts" without the need to transcribe tapes.

Check-sheets gave me the "numbers" part of my study, what colleagues would consider the "hard, unbiased" data. Mine was among the first quantitative studies of a wild primate. Yet even in my earliest days with the baboons, I realized the limitations as well as the benefits of check-sheets. The chief benefit was neutrality. Before joining the troop each day, I picked several individuals—male or female—to follow, and that way avoided unconscious preconceptions. Each sample animal's activity got logged, not just that of the loud or aggressive animals. But how long should each follow last? I tried different lengths, finally deciding to watch a "sample" baboon for 30 minutes before going on to the next one.

Yet there was no way the check-sheets could capture the wildly multifarious swirl of baboon activity on a typical day. So, I started a daily diary that elaborated on details of the group's life (see appendix 2). I noted where they slept, when they began their day, their movements throughout the day, and my best guesses at their reasons for doing what they did. I wrote about the little dramas: the stirs caused by a sexually receptive female or a newborn baby. And I described the big events, like the encounter with another wild troop and the predatory stalking and capture of a young gazelle.

The diary put flesh on the bones of the datasheets, which soon lengthened to capture categories of behavior I'd never read about nor expected to see. Of course, I recorded who groomed whom and when one baboon displaced another. But paying attention to their interactions helped me figure out why certain baboons chose to groom each other and why others avoided each other. The combination of check-sheet follows and the troop diary were giving me a "rich" set of information. They were also bringing the improbably captivating social life of baboons into focus for me for the first time.

Early on, I realized that the fascination with large males had produced a lopsided and possibly distorted picture of baboon society. I'm often asked how the earlier baboon studies made such a glaring procedural mistake, focusing as they did on those big males. The answer came to me later: Researchers then were blind to their bias. They followed accepted scientific practices of the time, and they were products of their gender-unequal era. Consider stereotypical human gender roles in the 1950s and early 1960s—breadwinner and housewife—and you'll sense how times and the zeitgeist can shape scientific interpretations, and also how far we've come. Science can't be hermetically sealed away from society. This was my first whiff of that principal.

And so, it seemed quite a few presuppositions about baboons were in need of deeper scrutiny. That included the baboon model I'd come to Africa to test, in my mind now a little ragged around the edges . . .

## Chapter 3

# The Model Breaks (1972–1981)

**FIELDWORK INVOLVES RULES,** and like everything else, rules evolve. In the 1960s, Jane Goodall broke the rules by naming "her" chimps, though doing so implied projecting human traits onto animals—anthropomorphism—a scientific sin at the time. Yet her study yielded important discoveries about chimp society. I needed the best way to keep track of baboon interactions, the very point of my study. By naming my subjects, I wasn't so much breaking a rule as making a necessary new scientific exception.

Then there was the rule about nonintervention. This I took seriously. I was there as a camera, to record all I saw and later to analyze and interpret the data I gleaned. Given my objective—to test the baboon model—there was little value to the effort if I wasn't watching something very close to natural baboon behavior. In the field, nonintervention can be tricky. Early on, for instance, shy Naomi's little daughter, Robin, often came too close and once even touched me.*

*Named before my rule took effect. Naomi's subsequent infants were named Nancy, Nigel, and Nevada.

Unlike her mother, who mostly kept to herself at the group's edge, Robin was curious and unafraid. This posed risks for me, both personal and professional. If I inadvertently scared her, adult baboons might threaten me. And if I were to engage with her—it was tempting—I would have to question the scientific integrity of my study, and myself.

I was lucky. A baboon "ethogram"—a list of baboon behaviors like grooming, approaches and avoids—had been started by Thelma Rowell and expanded later by Tim Ransom. When I was still finding my way, the ethogram provided a foundation to build on and basic insights into what male and female baboons usually do. I also had "permission" to spend several months just "getting to know" the animals. It was not yet okay to talk about animal personality, a taboo for another 40 years. This would not stop me from noticing individual personality "traits" as the animals grew familiar. But my focus was trained on describing two things: baboon actions and baboon reactions, those of both male and female, just the social—my means for testing the baboon model. Straightforward, but not simple. Baboon actions and reactions could swiftly ramp up like a beach town melee during spring break.

The males baffled me from the start, even before I began collecting data. They were seldom aggressive. This wasn't a fatal blow to the baboon model, since the function of a male dominance hierarchy was to convert what could be a perpetual "war of all against all" into a patterned and predictable set of male relationships. Still, I was mildly surprised that aside from the usual foraging and resting, a male's typical day involved mostly quiet socializing with females and infants, and maybe a few brief low-key interactions with other males.

Moreover, any sense of a fixed hierarchy in the male population felt elusive. In fact, male ranking struck me as surprisingly fluid. I had already taken note of a set of ritualistic-looking approaches and avoids. These I came to understand were "greetings" between males using the only language at their disposal, behavior and their bodies. Years later, I learned that one of the scientists behind the baboon model, Irven DeVore, saw so little male aggression that he stirred it up by tossing food to male baboons. Only by inducing competition could he determine who was dominant to whom. Might that have distorted male behavior? Seems so. Could it be that male aggression was the exception, not the rule? Possibly.

I thought about those "special" episodes when a couple of agitated males verged on tearing into each other, then one fetched a baby and suddenly it was over. Was male aggression more charade than threat? And what about that baby held up like a shield? Then epiphany, after months of observation: All hinged on whether the baby screamed, and at whom.

When a baby is brought into conflict, there are two possible outcomes. In one, things work out as they did for Carl: The aggressor backs off at first sight of the infant and the infant remains calm. In the other, the aggressor persists, causing the baby to scream in terror and all hell to break loose. Troop members typically avoid interfering in male interactions, but they loudly and vigorously defend babies and females in distress. In these instances, the troop fiercely "mobs" the aggressor, scolding with intense high-pitched calls that leave no doubt about what they mean: *Stop that immediately!* The group then pushes the offending male to the edge of the group, where he stays in exile for hours. With consequences like this—something that looked like a human scenario of crime and punishment—a male baboon quickly learns to back off when a rival brandishes an infant. Since I found "mobbing" upsetting, I imagined it was worse for a baboon.

But there's another wrinkle. The defending male can't recruit just any baby. If he makes a wrong choice, the infant may scream at *him* and the ploy will fail, sometimes spectacularly. Early in my study, Big Sam was a socially ignorant large adolescent, just learning the ropes. In a conflict situation, I saw him run off looking for an infant buffer. First, he grabbed a brown infant; too old, it had to be black. Then he reached for random black infants. They screamed and flailed because they didn't "know" him. He finally found an infant "friend," who joined him without a fuss. Lesson learned in a comedy of errors, one that alerted me to something else. Here was an early hint of the critical roles social bonds and friendship play in baboon society.

Recruiting an infant to subdue conflict now has a name: "agonistic buffering," a term first used in a study of Barbary macaques, but practiced differently in baboon society. I also saw baboon males use females as buffers, a variation unique to Pumphouse. No other studies have described it.

Male defense using infant or female buffers didn't happen every day. I saw it mostly during periodic upheavals in the baboon order, when the male hierarchy got wobbly. An analysis of the buffering episodes I'd tallied was revealing: Infant recruitment was a way "not to lose," with stalemate the objective. I scratched my head about that. The risk of injury in physical combat certainly made it better for the subordinate male not to lose. Meanwhile, the potential winner risked mobbing if he confronted a male with an infant buffer, so best to withdraw. Somehow the baboons had turned a win/lose situation into a draw that, paradoxically, benefited both, even if it might not have felt that way. Remarkable.

The structure of the male dominance hierarchy also defied expectations. No male held the top spot for any great length of time. For a few months, one male might look dominant, but rivals always waited in the wings. It might be a younger male coming into maturity, or a "new" male who'd recently joined the group, or even an alliance of lower ranking males. "Alpha" might avoid, or back away from, a rival one time, but to keep his position he'd need his rival to avoid *him* in subsequent encounters. When the day came that his avoids outnumbered those of his challengers, his reign in the top spot was over. The male hierarchy clearly *had* an alpha position, but it was filled on a rotating basis. I began to question the very idea of "alpha" male,* such a key element of the baboon model.[1]

The best clues for cracking the male system came from watching newcomer males, several of whom joined Pumphouse during this first study. Ray's arrival in particular shook things up. By way of introducing himself, Ray boldly approached and chased a number of Pumphouse males, often without provocation. Yet each male had a different response to Ray's threats. What would these varied responses tell me?

To find out, I began logging individual male's reactions to Ray's bullying. David, for one, retreated to immediate family, his safe place, and busily groomed his mother or a sibling, pretending not to notice Ray. Sherlock was ambivalent. Another large adolescent, Sherlock closely watched Ray, intrigued by his novel presence, but cautiously so. Sometimes he'd scoop up his baby sister and carry her on his belly, as a mother would, and move in close to Ray. With his sister as a "passport," he would settle into a front row seat at the Ray show. On his own, Sherlock would confidently stride toward Ray, then veer off at the last moment, a classic approach-avoid.† Ray could displace adult males like Rad or Carl individually, moving one or the other away from something he wanted, but not when they teamed up against him. During this first study, other males, like Strider, joined the group, and some, like Big Sam, achieved adulthood. Changes in the male sphere always brought disruption and rejiggering in the male hierarchy.

Constant shuffling in male order—the daily approaches and avoids, the challenges and varied responses—meant flux was more the norm than order. I could determine the degree of flux by

*Almost 20 years later, my observations about the "dynamic" male hierarchy were confirmed by the Amboseli Baboon Research Project.
†Later I followed Sherlock after he migrated to a new troop; his behavior always amused and instructed me.

counting male greetings. Since behavior is the "language" of baboons, I paid close attention to discrete behaviors, hoping to parse their meaning. It wasn't a simple matter. Baboons also need to interpret the behavior of troopmates, and meaning can vary depending on context, both social and spatial—that is, where, physically, they happened to be. Context became a data point for me, along with individual behaviors.*

The gestures and postures males use to greet each other evolved from sexual behavior: mounts, pelvic thrusts, and diddles (like it sounds, a male touches another's genitalia, perhaps assessing size). A male might narrow his eyes and shake his head from side to side, a gesture meaning "come hither." Sounds such as grunts, lipsmacks, and gecks (a human word for a baboon distress sound, drawn from ethograms) often punctuate these gestures.[2] While males rarely form close bonds, they do have an acute awareness of each other. I've observed males a quarter mile distant signaling each other. A male who initiated a greeting was not always dominant, nor was the male who ended a greeting. But dominance *could* be discerned in contexts other than greetings, such as a subordinate male turning away from or avoiding another male. When males in a troop engaged in more greetings than usual, it was another sign of churn in the male ranks. I think greetings helped the baboons in the same way they helped me, offering a sort of temperature check of the male hierarchy. Like me, they closely monitored everything happening around them.

As I watched, Pumphouse males seemed almost to be flouting the baboon model. They tested each other, jockeying for position and assuming different ranks at different times. Also contrary to the

*Context was a point of contention. Most researchers then considered context to be redundant of the nature of an interaction between two animals. Forty years on, most agree that context adds meaning in behavioral studies.

model: The top male wasn't always rewarded with the best of everything, not even sexually receptive females. Recall the baboon model's claim that the top male monopolized females and sired the next generation, thereby producing the "best" babies and the group's best shot at ongoing survival. What I saw did not always support this claim.

Female baboons have a 40-day menstrual cycle.[3] Ovulation happens mid-cycle, when her sexual skin is fully engorged. Males often poke and sniff, seeking scented pheromone cues as well. At the right moment, a baboon male forms a temporary liaison called a consort with a sexually receptive female. However, the consort couple is seldom alone. Other interested males follow behind the consort pair, trying to "turnover" control of the female. Sometimes a dazed infant tries to keep up with its mother as she is wooed but gives up and seeks comfort from relatives or playmates. Juvenile males often harass a consort couple, but they seldom succeed in interrupting a copulation.*

But there's a key ingredient in a successful mating, one that took me by surprise. The receptive female must be willing.[4] If she's not, even the most dominant male will face rejection. In other words, male rank is no guarantee of female willingness, and willingness is essential to success. As with most primates—humans being an unfortunate exception—rape does not happen in baboon society. Successful copulation involves several pelvic thrusts, an ejaculatory pause, and, most important, a cooperative female. After seeing a number of failed copulations, I fully grasped the power of female cooperation. A dominant male could be twice the female's size, but if she demurred, size and rank would not get him the most vital resource, the chance to produce babies.

*Fun fact: Male baboons prefer mature females to newly adult females, though it seemed to me the young ones were more attractive. I have yet to learn what counts as attractive to a baboon.

It was obvious when a sexually receptive female was not interested in a particular male. She would run him ragged with evasive feints and dodges and dart away from attempted groomings and copulations. Eventually he would concede defeat. She would then seek out a favored male, one she knows and likes, to initiate a new consort, or wait for a preferred male to step up. In either case, it seemed a display of something resembling agency. Female cooperation now has a name: "female choice," the baboon equivalent to "consent."

Female choice doomed dominant newcomer Ray to celibacy in his early months at Pumphouse.* He'd quickly ascended in the male hierarchy, not only by aggressively intimidating other males, but also with his attempts to steal receptive females from their consorts. Yet in those early months, I never saw him complete a copulation. His advances were spurned time after time, by female after female, as rivals hovered nearby. Poor Ray almost radiated anxiety. Even his attempted groomings failed. Given his dominance, the baboon model would have predicted success for Ray, but he was a reproductive flop until he'd established himself in the group.

I also saw older males in physical decline play the mating game better than their younger, stronger counterparts. One way was through a maneuver I called "sidelines." Sumner, an older male and close friend of Peggy, was a master of sidelining. Unless you were following him, as I was, you'd have no idea he even noticed a consort pair. He would seat himself behind a bush and discreetly monitor a cluster of males vying for a receptive female. He knew tension between the consort male and his challengers would likely escalate to aggressive bluffing. That would be his cue. As the males mixed it up, Sumner made his move, skillfully steering the female

*By virtue of their unfamiliarity and unpredictability, newcomer males start out as dominant in a group. Which is not to say all newcomer males strike out. Females sometimes find their novelty appealing.

away from the others and finishing the job. Sumner also had another tactic. First, he would join the males jostling around the consort couple. Then, with a few subtle moves, he'd instigate a mobbing of the consort male. In the ensuing fracas, Sumner was nowhere to be seen. As in sidelines, Sumner exploited chaos to claim the female, copulate, and walk her away. Only a smart male could do this, one who knew how to read a situation and predict the outcome of an action. Sumner might have been in his late twenties, in no shape to challenge young rivals with physical strength. But he wasn't out of the running when it came to making babies.

I was beginning to think of the evolutionary concept of "fitness," then defined as reproductive rewards based on physical strength, as narrow and limiting.[5] Shouldn't qualities like ingenuity and strategic behavior also be taken into account? These were early days, before serious discussion of animal "mind" took off, but my observations had already launched me in this direction; I had seen baboons use social strategies *instead of* aggression in competition and defense. They were convincing me—and soon others—that baboons, at least, have "mind" and can think tactically (see chapter 14).

The males were coming into focus as my cache of data grew. So far, during my first year in the field, I had seen a fluid male hierarchy and not much male aggression. I had observed tactics like agonistic buffering that specifically averted conflict, as well as other ingenious alternatives to competition for a critical resource: receptive females.[6] A striking pattern was emerging from these interactions, and it involved relationships. I'd been wondering about why individual males seemed to seek out and consort with certain females. Then I made note of successful copulations and observed how these consort couples related in day-to-day life. That didn't tell me much about the appeal of particular females, but I did discover that females often "chose" males they'd already bonded with. A bonded female kept close to her male consort, allowing him to focus fully on fending off rivals. Her loyalty made his life easier and an overturned consort less likely. It made sense that males worked to create social bonds with females.

Not only did males forge relationships with females; they also cultivated bonds with infants. Tim Ransom had seen these behaviors at Gombe Stream Reserve in the 1960s and called them "special relationships," but he did not plumb their significance.[7] I thought I had the answer. It seemed clear to me that both copulations and agonistic buffering, when successful, hinged on bonds created by these "special relationships." From these bonds grew trust, as important to baboons as to us.[8] That meant that the first order of business for a newcomer male was to convince at least one female and her family that he wasn't unpredictably aggressive and could be trusted to look out for them. In the case of agonistic buffering, as we saw with Big Sam, a defending male first needed to gain the trust of an infant.[9] If the infant he grabbed did not know and trust him, it would scream in fright and bring the troop's wrath down on *him*, not the aggressor. Baboons, like us, need to *earn* trust, and male baboons earn that trust when they cultivate relationships with females and infants. We will soon take a closer look at that process.

---

Pieces were coming together, a picture was forming, and in that picture were many things the baboon model had missed. I'd become convinced baboons had options, choices to make about which of several possible things to do. And it wasn't that "might makes right" or "winner takes all," as the baboon model suggested, because aggression was only one of the tools baboon males used in competition and defense. I decided to use the term "social strategies" to describe these behaviors.[10] In most cases, these strategies got better results than ferocious bluster. It made sense to me, and it aligned with evolutionary principles. Certainly, a male baboon would opt to use his wits instead of might, if possible, because aggression is risky. Injury can incapacitate a male for a long time. That meant successful males were smart, skilled, and strategic.

These skills and behaviors gave every appearance of being calculated. But how could that be? We thought humans were the only animals capable of strategic behavior and intentionality. Yet I'd watched as baboons deployed strategic intelligence and cultivated trusting relationships. I saw what looked like human negotiation as baboons traded favors and benefits like grooming and mating. Special relationships and social strategies explained away most of the list of unexpected behaviors I had labeled as "exceptions" to the baboon model. They weren't exceptions; they were routine. I realized they also punched a hole in the model. I couldn't deny that evolution had built male baboons as fighting machines, and yet real fights were rare.

Females, not just males, fit my new view of baboons as strategists. I already had hints that females played a more central role in monkey society than the baboon model suggested. High-ranking females like Peggy and lower ones like Naomi maintained their standing no matter which males—high or low—associated with them. That meant that females had their own dominance hierarchy, and that it was firmly fixed. A similar finding came out of research in the late 1960s on the female hierarchy among rhesus monkeys on the Caribbean island of Cayo Santiago. Glenn Hausfater, who was tracking yellow baboons in Kenya's Amboseli National Park when I started my study, also found a stable ranking system among females. Female status, unlike that of males, followed family lines, meaning baboon females had a de facto class system. Mothers of high rank give birth to high-ranking infants, and so on down the hierarchy. In contrast to the dynamic male system, the female hierarchy looked stable and predictable. Decades after my first study, when I watched descendants of these original baboons, the relative ranking of females by family persisted, despite some matrilines having died out.

But rank could not determine every outcome. Females also needed strategies to accomplish what their smaller bodies could not, especially since any large male outranked any female. Baboons love meat, and my favorite female, Peggy, was no exception. But females rarely capture large prey. A young adult male named Dr. Bob had caught a young Thomson's gazelle, called a tommie. Others caught wind of his prize and moved in, hoping for a share. But Dr. Bob took off with his quarry, trailing scraps that served to distract his pursuers. Then along came Peggy. She and Dr. Bob had a special bond, so he gladly let her sit down and begin a grooming session.[11]

Grooming's importance to baboons can't be overstated. It's a blissful activity that also creates and deepens bonds between animals (excepting adult males, who rarely groom each other). Equally important, it's a social tool that can be bartered. Gently and methodically, the groomer parts and sifts through its companion's hair, removing bits of dirt, burrs, and insects, some of which they pop into their mouths.* Grooming produces a sort of hypnotic state of calm, with eyes closed and movement stilled. Primates, including humans, are wired to be pacified by touch, which explains massage and jetted hot tubs in our society.

Peggy's grooming worked its magic on Dr. Bob. Lulled into oblivion, his hold on the carcass went limp. Peggy made her move. She snatched the tommie from Dr. Bob's lap and ran off, dragging the carcass behind her. Peggy ate her fill while others claimed the bits she'd dropped. Smart Peggy. However, several days later, Dr. Bob got another tommie. I watched the same scenario unfold, but this time Dr. Bob was prepared. The moment Peggy paused her grooming, his hand came down on the carcass. Smart Dr. Bob. This time, Peggy just got his leftovers. I observed similar sequences with different

*Animal trappers have said that baboons were the cleanest of the wildlife they captured.

baboons and marveled at the clever tactics both female and male baboons used to outsmart each other.

I'd seen nothing in the literature about social strategies or the different ways males and females manage their social relationships. Nor had anyone previously studying baboons seriously considered the female hierarchy,* which was looking to me like a key factor in the group's structure.[12] Females are conservative: They seldom leave the group, rarely contest their hierarchy, and maintain a sort of feudal system, all in striking contrast with the continuous slow boil of the male system. How was it that these two distinct systems could coexist in a single cohesive baboon group?

Mating, of course, brought them together, and immatures played together. Both males and females shared an attraction to little, black-coated newborns. But male-female and infant-male bonds were the strongest glue of all. I was seeing the complex negotiations these bonds required. Each time a male left or a new male arrived, new networks of relationships had to be built.

Negotiation is a term casually used by humans to mean a dialogue between individuals that is meant, in the end, to satisfy both.[13] By definition, it's a transaction that signifies an arrangement.[14] People negotiate in daily life, naturally and without much thought. I think baboons engage in the same sort of give and take, but without the advantages of language, symbols, or material culture. Instead, they "talk" using bodies, gestures, and sounds. Groomings are offered and help provided when conflict flares. Help may be withheld to express displeasure. Lipsmacks, grunts, and facial expressions carry meaning as well.

I was troubled from the start by which *words* to use to describe these new baboon insights. Per my ethological training, I had been cautious about using "human" words like "friendship" to

*But Glenn Hausfater studying yellow baboons in Amboseli came to the same conclusion.

describe baboon special relationships. At the time, the specter of anthropomorphism—the projection of human traits on animals—loomed large as a foe of scientific objectivity. I struggled with this problem for many years. However, I could plainly see that baboons had many humanlike traits. This made me more accepting of words usually reserved for people, words like "friendship," because the longer I watched, the more baboon "special relationships" looked like human friendships. Baboons choose their friends and spend most of their time with them, meanwhile paying little attention to nonfriends, just like humans. Baboon friends spent so much time together that I came to think of them as "behavioral kin," because they created intimate and intense social networks normally seen only among real, biological relatives.

I saw a social network being built by Ray, the immigrant male initially shunned by females. First, he had to identify existing friend groups, which he did as I did: by observing the behaviors and interactions of his new troopmates. Then he focused his attention on a single female, Naomi, Robin's mother. Since all new males are unknown quantities, and therefore frightening, getting close to Naomi and her family was a painstaking process that required patience and persistence. He began with gentle grunts and lipsmacks and hung around nearby without causing trouble, showing he could be trusted. The breakthrough came when Naomi allowed him to groom her. Acceptance by her family and friends followed. Ray was on a roll. His next target was highest-ranking Peggy.

Friends rested, slept, and foraged together. They groomed each other. And, crucially, friends supported each other when conflict flared. Of course, a big male's support was objectively more advantageous than that of a female or infant. But females could be "friends with benefits" and infants could help deflect violence.[15] Friends, no matter the size or sex, had each other's backs. Yet friendships were not unconditional. I saw friendships end as well as begin.

As the baboons changed my mind about their society, I grew comfortable using words like negotiation and trust, though they'd once sent up red flags. The social glue of friendships was trust, and trust grew from mutually helpful interactions between individuals, what I now consider negotiations. Friendship was a behavioral quid pro quo that promised support and benefits for both animals, and bonded them. To what end? The bigger picture—the evolutionary picture—suggested that baboons' social strategies minimized harm in existential situations involving competition and defense. That is, the social* appeared to underpin survival.[16]

So, the original "baboon model" was falling short. It might have passed scrutiny as a snapshot, but it was proving hopelessly reductionist the deeper I went. The reality of a baboon troop was messy and variegated, a socially *complex* thing, less likely to conform to a model than defy it. For a male baboon, what mattered wasn't just fighting ability, but the social smarts that come with age, experience, and membership in a group. The model didn't address the need to acquire social skills or describe what they looked like, and it said nothing about the building of relationships. Why was that? Science, including the science of baboon behavior, starts with simplifying assumptions. The scientists who extrapolated the model made the mistake of thinking the baboons' system was simple, but it wasn't. I was learning that lesson.

Improbable as it seemed, it looked to me like baboons were practicing a version of the classic Golden Rule. The formation of social alternatives to aggression *required* baboons to be nice to each other. They treated others as they wished to be treated, using bodies and behavior in an elaborate system of negotiated reciprocity. In doing so, members of baboon society also connected with individuals outside the circle of kin and family, establishing social capital

*And now studies have shown that female baboons who are the most social live longer.

through reciprocity and exchanges among unrelated friends.[17] Social ties are key because they build social capital and give baboons options. An animal with social capital can *choose* to call on trusted friends to help resolve conflict and avoid physical aggression.

I began to think I was starting to see the world through baboon eyes. I became deeply attuned, as they were, to subtle gestures and shifting contexts in their social world. For example, an infant becoming an agonistic buffer didn't "just happen." The adult male first needed to understand the tactic, then cultivate a relationship that would assure an infant's willing cooperation. From Descartes to "modern" animal studies, animals, including baboons, were assumed to be stimulus-response machines with little or no volition or agency.[18] But behaviors like agonistic buffering, consort strategies, and female choice had persuaded me that baboons were more than that. I called my first book *Almost Human* because these early findings convinced me that baboons were much more like humans than anyone supposed.[19]

My own model of baboon society was coming together piece by fascinating piece. It seemed my brain was becoming a sort of baboon behavior recognition machine. Scientific discovery is as satisfying as solving any type of mystery, and baboons had proved far more interesting, and mysterious, than I'd expected. For each question I answered, a new one popped up. Uncoupling the tight theoretical link between male dominance rank and male evolutionary payoffs exposed the earliest evidence of baboon social *complexity* . . . as well as the fragility of the baboon model. And I'd only just begun.

Darwin was impressed by what he knew about baboons, which was not much beyond the anecdotal. In *The Descent of Man* he wrote admiringly of "that old baboon, who, descending from the mountains, carried away in triumph his young comrade from a crowd of astonished dogs."[20] But Darwin never attempted to explain *why* the old male did what he did. This omission was the basis of what I came to think of as Darwin's monkey puzzle. I now had some of the pieces. Darwin had not seen what I had, that baboons have a society ordered by negotiation, reciprocity, and trust, a society in which friendship matters. In other words, they look out for each other as they would have others look out for them.

At the end of my first baboon study, in 1974, the University of Nairobi invited me to speak about my research to a small group of researchers and students. It turned out to be a momentous occasion, and not because the talk was a sensation, though it went well. Fate would have it that someone in the room, another speaker that day, was destined to play a central role in my life. Jonah Western, who I knew of but hadn't met, had been studying the destruction of the Amboseli woodlands since 1967. His insights about who or what was to blame had put him at the cutting edge of conservationist thinking. I was a bit dazzled by his good looks, and maybe by his British accent (though he was raised in Tanzania). I also liked what he said and the way he thought. Perhaps I recognized a kindred rebel spirit. In any case, then was not our time, and would not be for nearly another decade, though we would cross paths professionally and socially from time to time. I think we both sensed something, even at that first meeting in Nairobi.

For my doctoral thesis, I'd set out to study the roles played by males and females in baboon society as a means to test the baboon model. The data I had so diligently gathered was coded onto old-style punch cards and fed into Berkeley's mainframe computing system, a beast occupying eight floors of a large campus building. But the computer crashed, mangling the cards and ending any hope of analysis. Washburn suggested I pivot to a tangential subject: my observations of baboons hunting gazelles as a group. It was something never seen before, and excitingly, looked like the invention of a new baboon "tradition." Oddly, this serendipitous shift conferred a sort of celebrity status when my findings got out. Conference invitations poured in and a leading magazine, *Science*, published my research.[21] *National Geographic* put me on its cover and wrote a splashy story that proved a mixed blessing.[22] Meanwhile, on Washburn's recommendation, I had secured a teaching post at the University of California, San Diego. They agreed to split my year between classroom and field. Things were going well after all.

Still taking shape were my ideas about the unstable male dominance hierarchy, the social strategies baboons use in competition and defense, and how smart these animals had to be. Today, decades later, many of my early insights are treated as facts and taken for granted. Contemporary baboon studies freely discuss social complexity, baboon politics, and the baboon's thinking mind. Former skeptics now mostly agree that baboons favor collaboration over aggression. At the time, however, these notions were nothing short of heretical. When I ventured to tell Washburn about "new" baboon behaviors I'd seen with my own eyes, he, yet again, shut me down. The world of primate studies, devoted as it was to the study of evolution, was feeling curiously unevolved to me. So, I mostly kept these ideas to myself in those early years. I had a career to build, and the baboon model was still gospel in the field, even coloring the popular perception of baboons. The transformation of my thoughts into a convincing argument—representing a more complete and infinitely more nuanced baboon model—was to take decades.

I knew there was more to find out about baboons, but I was in no hurry. Years passed and my fieldwork continued, because my fascination with baboons turned out to be boundless. I thought I had all the time in the world. Then came an event that tested both my scientific imagination and my ethical framework.

Chapter 4

# Bitter Harvest (1981–1984)

I HAD BECOME the baboons' most constant watcher. Nine years passed, and I started to feel I knew them well. Perhaps not many surprises remained, I thought. I was wrong. In 1979, the owners of Kekopey Ranch sold the land. The idyllic life I shared with the baboons ended abruptly and veered into the unknown. It's a story I told in my first book, *Almost Human*, but I retell it here with the illuminating benefit of hindsight.[1] Turns out the nine years I'd spent with the baboons were just early chapters in a quite different book about what it means to be a baboon in today's world.

In 1979, the biodiversity crisis had barely surfaced in our collective imagination, though alarms were beginning to sound.[2] Even reputable scientists were suddenly invoking New Age warnings: Harm to any part of Gaia, Mother Earth and nurturer, would damage the whole. The baboons and I found ourselves on the cutting edge of change. Habitat loss and its threat to animal populations weren't fully formed "issues," but we were living them. The sale was our tipping point. For the first time, but not the last, the baboons would show themselves to be heralds of an altered future.

Kekopey had become a pawn in Kenyan politics. The white "settler" family had acquired it "legitimately," when the British colonial government annexed much of Maasailand and deeded it to colonists. After Kenya gained independence in 1963, its first president, Jomo Kenyatta, recognized all preexisting title deeds. But by the 1970s, he wanted land ownership to be "Africanized." Under this program, if Kekopey's owners sold to a Kenyan farming cooperative, they could take the proceeds out of the country. They sold, of course, and moved to England.

Transformations unfolding in the late 1970s were modest compared to what was to come, but were still a sea change for my baboon study. Rich agricultural land was in short supply. Over generations, a Kenyan family's single plot would be subdivided many times, resulting in parcels so small that descendants were unable to support themselves. The government's program was designed to help.

After the sale, small farms (*shambas* in Swahili) popped up on Kekopey. Newly arrived farmers had their work cut out for them in this rangeland. Where livestock and wildlife had thrived, rain-fed crops needed coaxing. Before long, a cluster of shambas had sprouted in the baboon range. Soon crops would be growing on land where the baboons once found natural foods. Then what would happen?

Before the first harvests, baboons traveled and foraged as usual, only now they had new playthings. Baboon youngsters have many games. In a version of hide-and-seek, one races up a tree, then drops down on an unsuspecting playmate. Now I watched as a team of immature males, led by Bruno and Berlioz, discovered that the thatched roof of a farmer's hut made a great slide. Plastic containers, discarded jars and cans, and a Coke bottle served as excellent new toys. I laughed, but the reality was that baboons were wreaking havoc on a homestead. I didn't want to think what would happen when the owner returned.

Soon several large males discovered dried maize and beans stored in a hut. They quickly ate their fill. These were early baboon forays onto farms. With luck, the baboons would turn out to be mere nuisances, damaging huts and snacking on stored seed. But knowing baboons, I feared crop raiding could become a habit and an issue. I'd heard rumors from elsewhere in Kenya about baboons invading small farms and decimating crops, and I found them plausible. If it happened here, we would be in a painful new reality with a problematic new X factor . . . hostile people.

My dread was affirmed. The baboons noticed when the first crops sprouted. Anxiously, I watched as the males of Pumphouse entered a field and inspected this new resource. They liked what they saw. It didn't matter that these crops were immature. The baboons ripped the plants apart, ate what looked good, and left destruction in their wake. A disaster. What could be done?

I'd maintained my nonintervention policy and my scientific stance. What I *could* do was turn disaster into an opportunity by proposing a study. Crop raiding had never been studied, so there were no hard data suggesting ways to prevent or deter crop raiding by any nonhuman primate. I saw a need to address the human element as well, in hopes of minimizing conflict with the animals. Up to this point, I'd focused only on the baboons' social behavior. A study of raiding would need to include baboon feeding and ecology. Of course, after spending nearly a decade with the baboons, I knew something about their feeding habits. Earlier studies had already concluded that baboons are "opportunistic generalists," with eclectic tastes that made it easier to list what baboons did *not* eat than what they did.

Take baboon hands. Baboons have primate hands that deftly harvest a wide range of foods. I saw this every day. Corms, for example, are the underground storage organs of sedges, a grasslike plant, and they're an important part of the baboon diet in the dry season. But they are accessible only to animals smart enough to know of them and handy enough to extract them from the soil. I watched Robin, now an adult female, as she prepared a corming site. First, she pulled away the part of the plant that was above the ground. With that area clear, she dug several inches down to reach the corms. Then, having pried out the small onion-like corms, she brushed off residual dirt and popped them into her mouth. This was a process beyond the ability of a gazelle or cow and denied those animals a nutritious dry-season food.

My understanding of baboon feeding decisions deepened when I documented the seeming genesis of a hunting tradition, initiated by Pumphouse males. (This finding became the accidental subject of my doctoral thesis.) I'd seen baboons prey on young Thomson's gazelles—tommies—but this was different. Tommies' defense against the large cats that hunt them is to freeze in place in the tall grass. This makes them invisible to the cats, who rely on motion, not color, to spot prey. Baboons, with their color vision and depth perception, easily see a "hidden" tommie and move in to "collect" it. Typically, a single male stalked and captured the creature.

The new style of predation began with Rad, the first male I learned to identify. I'd noticed Rad leaving the troop one day and followed him to a nearby herd of Thomson's gazelles. As I watched, he isolated a baby from the herd and chased it down. Several times, I saw Rad repeat this practice, always out of sight of the troop. But when Rad returned to the troop with dried blood on his arms and face, others took notice. Soon Carl and Big Sam joined Rad and together they discovered a predation technique that cost the lives of many young gazelles. I was gobsmacked when I first saw these males chasing a tommie toward each other in relay fashion, not randomly.[3] Up to this point, chimpanzees were the only primates seen to hunt collectively, besides humans. It was an exciting development, and it shook things up in the world of primate studies, at least for a while.

But raiding was not the same as corming or hunting live prey. Whatever insights I'd gleaned from observing those activities were not going to explain raiding behavior. Once again, I found myself in terra incognita, poised to be the first to study an unstudied behavior and lacking a road map. But this is what makes science fun, and maybe a little addictive. There's always something new to learn, and the chance of a fresh discovery. Drawing on my history with baboons, I framed the raiding study proposal as a means to find out how baboons and people would react to each other's activities in these new conditions, and also to deal with problems that might arise. A number of granting agencies, including the World Wildlife Fund, the New York Zoological Society, and the Leakey Foundation, approved.

---

**September 1980**: Sunrise. When skies are clear, daytime heat escapes overnight and mornings feel numbingly cold in the shadow of the cliffs, especially when you don't have a modern puffy jacket. Now new shambas covered about a tenth of Kekopey, but were concentrated in a few places that, unfortunately, were critical areas for the baboons. The baboon sleeping site in front of me has two special features: a fig tree as tall as the fault scarp, and a lovely small grassy field nestled halfway up, just to the north of the fig. My climb to the top is a taxing 100 feet straight up, but the amazing views and relaxed baboons make this a favorite spot.

Baboons concentrated on grasses and sedges here in "Happy Valley" because Kekopey's savanna had few trees, mostly whistling thorn acacias which grew in isolated stands in sticky clay soil called "black cotton." These stubby trees whistled when the wind swept past holes created by the ants that lived in the trees' numerous swollen thorns. Baboons gnawed around the thorns to get at the ants and made a sound like children popping bubble gum.

When I began my crop-raiding study, I was confident of one thing only: I already knew as much about baboons as anyone else. Yet watching the new raiding behavior was proving harder than I'd expected. No longer did I bask in the scenic beauty of corming sites near rock escarpments. Now I was hanging around small farms waiting for baboons to attack the fields. A few baboons discovered edible items in human garbage, so refuse pits became a new attraction as well.

**September 1981**: A year later. Cold again, standing in the shadows of the baboon sleeping site, near the large fig tree. The baboons loped to the top of the cliff, followed by this less agile human. When I reached the top, the group was resting, grooming, and playing. All seemed normal until the negotiations began. On a typical morning, certain individuals will influence the direction the group takes in the search for food. A young male will often dart out front, pretending to lead, but he lacks the sort of influence that would actually mobilize the rest of the group. Then, when the troop veers away from him, he rushes to the front as if he'd led them in this new direction. Later, I would learn how remarkably complex troop movements can be (see chapter 8), but for now my focus was fixed on whether the baboons would raid.

This morning, a cohort of baboons who'd had a taste of raiding tried to draw the whole troop toward the shambas. Chumley, a large adolescent male, "indicated" the direction he wanted the others to take: He ran out front, set himself down, turned to face the others, and beckoned with grunts. Duncan, another young male, joined Chumley; Higgins followed. These three were already on their way to habitual raiding. I held my breath. Would the whole troop agree to follow them to the farms? No. Matriarch Peggy firmly refused and convinced her large family and friends—both male and female—not to follow. Like a rubber band pulled too taut, baboons who'd seemed tempted pulled abruptly away from the three males.

The raiders gave up and went by themselves. I followed them. We neared the shambas, with me nervously trailing behind. Seeing this, a farmer might assume I was herding the baboons like livestock, and he could blame me for the crops taken. Today, no one was home at the first shamba. The raider males walked up and down rows of maize, checking each cob. I adhered to my nonintervention policy and only observed. But I did lie low in case someone came out of the hut. Typically, there would be people, and dogs as well, but not this time. The raiders uprooted young plants, ate the tender new parts, and discarded the rest. Decades later, I saw a small herd of elephants decimate five acres of crops in just a few minutes. Baboons took longer, and were more methodical, but the result was the same: a season's food crop destroyed.

The baboons and I moved to the next farm, just a few hundred meters from the first. This shamba had beans, not maize. Beans are smaller and baboons have to work harder to harvest them. Even so, cultivated beans are bigger, more nutritious, and more accessible than the typical natural food. Oddly, there was no human presence at this shamba either. I'd already asked farmers during meetings and door-to-door visits to please chase the baboons away, in the hope that human vigilance would suffice as a determent. Sometimes it worked, but this morning the males entered the field unchallenged. That was good for me—no angry people—but bad for the raiding baboons whose taste for human foods was growing stronger by the day.

---

It surprised me that the Pumphouse raiders met with an equal and opposite reaction from the rest of the troop, which remained stubbornly resistant. Most days, the main group foraged on natural foods while the raider subgroup lit out for fields and homesteads. Then, over a period of six months, the raiding subgroup formed itself into a new fully fledged troop.* We called it Wabaya, Swahili for "Bad Guys." It was smaller than most, but what made it strik-

*Later I would realize that the splitting had been going on for some time (see chapters 9 and 13).

ingly distinctive was its composition. Only young adult and adolescent males and females were Bad Guys, though an older female, Beth, joined later. No youngsters or oldsters except Beth. Their social behavior was equally odd: Everyone groomed everyone else, not just kin or friends, the normal pattern.

I had preliminary findings. One was that only males of a certain age became raiders. Why was that? Raider males, compared to those older or younger, had "raging hormones," specifically testosterone. This made them both bold and attentive to nearby troops, often a prelude to switching troops. Not only were they adventurous, but they were also less attached to Pumphouse. Chumley, Higgins, and Duncan fit the profile perfectly and were, in fact, recent arrivals from a nearby troop.

Fewer females became raiders: Robin and Nancy, daughters of Naomi; Deidra and Dawn, David's sisters; sisters Melissa and Mavis; and cousins Desiree and Zilla. They shared two characteristics. One was that they were friends of one of the male raiders. The other was birth order. Each pair was the eldest and next eldest sister in their family.* That meant they ranked lowest in the family's hierarchy. In the baboon scheme of things, the matriarch is at the top of the family hierarchy. Each new infant she bears becomes next highest, displacing the formerly youngest sibling, and so on up the sibling chain. This is because mother always supports a younger child against an older child (see chapter 10).

A group will split when it is very large.[4] Consensus among primate scientists at the time was that groups break apart when their daily movements lose cohesion, at around 120–140 members. Families stay together but sort themselves into one or the other of two groups after the split. The Wabaya fission was unusual because it violated these rules. Pumphouse, the original group, wasn't

*Cousins Desiree and Zilla were the oldest daughters in their respective families.

excessively large at 90 members, and I didn't see coordination problems aside from the decision to raid or not. Moreover, families were fracturing. Most Pumphouse matrilines remained intact, even with some elder daughters leaving to join male raider friends.

Late in the fission process, Beth, an older female, defied expectations by joining Wabaya and turning raider. Beth was a favorite of mine. She was fluffy, not sleek, with "gray" silver hair that ended in light rather than dark tips. Her floppy tail was distinctive, resembling a long handle coming straight off her back, then curling down toward her legs. I'd known her from the start of my baboon work. Beth's family ranked the lowest in Pumphouse, but she was buffered by older sons who had grown big enough to be dominant to all females. She was wise in day-to-day matters, and she stood her ground, not deferring, as higher-ranking females greeted her newborns. Thanks to her husky sons, smart tactics, and male friends, Beth escaped many of the downsides of low rank. So I was puzzled. Why did she join the Bad Guys? She didn't fit the female pattern since she wasn't friends with any of the raiders, male or female. And she was a matriarch, not an elder sister in a family. Beth would not have made this move lightly. Of course, I could not read her mind, but I could watch her deliberations. She went with the raiders for a few days or a week, then returned to Pumphouse, then drifted back to the raiders for a spell before yet another Pumphouse return. Finally, Wabaya, the new group, won. She was its lowest-ranking female, but a raider nonetheless.

What might have convinced Beth to abandon friends and extended family? Clearly, crops were an attraction. Human foods are large, tasty packages of easy-to-digest calories compared to baboon natural foods. When later I had occasion to weigh Beth (see chapter 6), she was heavier than non-raider females. More astonishing was her reproductive renaissance. As an old female, she was having fewer babies. But after joining Wabaya, her fertility rebounded and resulted in several new births. She was still the lowest-ranking female, but Wabaya was small so there were fewer females above her. Could it be her new group offered relief from the stress of low rank?

Beth's infant, Billy, also surprised me. Billy closely resembled his mother. He had Beth's silvery color, fluffy hair, and a miniature version of his mother's "handle" tail. Billy was weaned, but Beth sometimes let him on the nipple and even carried him, though he was a big "brown" infant. They seemed tightly bonded, as often happens with infants of older females who have longer intervals between babies. At first, Billy moved back and forth between troops with Beth. But when she finally committed to the raiders, Billy stayed in Pumphouse. I found him there with his older brothers Benjy, Berlioz, and Bruno. My guess is that Billy stayed because he had no playmates in Wabaya, yet losing his mother seemed a big price to pay.

Beth's family saga does not end here. Eventually, Beth's only daughter, an adolescent, followed her mother into the raiding group. I supposed she shared her mother's reasons: access to human food and fewer females above her in rank. This made Beth the second-lowest-ranking female in the raiding group, just above her daughter. Again, I had to wonder if stress caused by low rank was the impetus to seek a new group with fewer higher-ranking females.

I deduced that food, friends, and dominance rank attracted female baboons to the raiding group. Were there different factors for the males, aside from youth and hormones? Large adolescent and young adult male newcomers to Pumphouse, like Duncan, Chumley, and Higgins, were experiencing growth spurts. Like the raider females, they were heavier than their non-raider counterparts when we weighed them. Though this benefit would not have been in their game plan, I had to wonder what advantages might accrue to them from added bulk.

I considered raiding in light of what I'd learned about male baboons as social creatures. All males appear to follow the same trajectory of social development, whether newly matured individuals in their natal group or newcomer immigrants. A newly adult male is socially awkward with few social skills. I'd also found that newcomer males were likelier to resort to aggression than the group's settled "resident" males. Once "friended," they relax. Whether newly adult or a newcomer, a baboon male needs social acumen both to make friends and to learn how to refrain from aggression, except in situations where social strategies won't work. When aggressive face-offs do occur, the winner is usually the younger, larger male, whose bushy mantle flares more menacingly.[5]

As a male baboon ages, social strategies become even more important, allowing him to compete and defend when younger animals challenge him. Since aggression is a tool more often used by young male baboons and newcomers, the additional body weight provided by human food offers a distinct, if unplanned, advantage.[6] While I'm not suggesting baboons consciously connect calorie-rich food with enhanced aggression, I do think evolutionary forces are at play. Evolution should have equipped these males to seek out "better" food, and in this case, it happened to be in farmers' fields. They were beneficiaries of a new and novel resource, but to them field crops were simply something new to eat, a welcome dietary add-on.

---

The answer to the question, "Why do baboons raid?" has changed in the decades since my crop-raiding study. Back then, primate scientists considered raiding to be "aberrant." This attribution bothered me from the start. What was aberrant about it? Farms offered ready-made food, and baboons need to eat. Of course, it was painful to watch baboons raid and even more so to deal with the consequences. But from a science perspective, I saw raiding as an expression of species potential, what an animal could become, and not at all aberrant. Would *any* addition to the baboon bag of tricks be aberrant?

I also used a revamped evolutionary agenda to help me interpret baboon raiding. I'd been trained in ethology, an evolutionary framework pioneered in the 1920s and 1930s and the foundation of modern studies of animal behavior. Karl von Frisch, Konrad Lorenz, and Niko Tinbergen won the Nobel Prize in Physiology or Medicine in 1973 for their discoveries about animal behavior.[7] Their work posited links between physiological and behavioral mechanisms in the development of individual and social behaviors. These founders of ethology were reacting against the overly humanized interpretations of animals that appeared after science accepted Charles Darwin's theory of evolution by natural selection.[8] In Darwin's new tree of life, humans were closely related to other primates and, for a time, anthropomorphism came into vogue.

Ethology put studies of animal behavior on a firm scientific footing by adopting new methods that corrected for human bias in interpretations of animal behavior. There were few primate studies, either in captivity or in the wild, before World War II. The war interrupted all field studies. After the war, Washburn, my mentor, initiated a round of new primate field studies.* In most cases, these were the first scientific investigations of a particular species in the wild.[9] The studies concluded that, in line with the "baboon model," all primates share the same basic behavioral patterns, with slight variations attributed to geographic separation and ecology. Individual animals in groups could be described by characteristics and "roles," a concept borrowed from social anthropology. Roles included "primate male," "primate female," and "primate immature."

Yet, as primate field studies resumed, they revealed an unexpected degree of variation among different primate groups. Why were langur monkeys in one part of India so different from langur monkeys in another part? Why did bonnet macaque and rhesus

*Summarized in two volumes.

macaque monkeys behave so differently, despite being closely related? Explanations at the time relied on two different theories. In one, differences came from adaptation to different ecologies.[10] The other theory proposed that evolutionary history, "phylogeny," was the answer: Isolated groups had simply evolved in distinctive ways.[11] Neither theory could explain the degree of variation that new field studies found. For example, the great apes—chimpanzees, bonobos, gorillas, and orangutans—are all closely related and clearly share an evolutionary history, yet their social ways range from almost solitary (orangs) to big communities (chimps). Chimp, gorilla, and to some extent bonobo habitats overlap, yet these primates have varied social organizations.

Previously, anthropologists felt that the study of nonhuman primate behavior would reveal overarching primate characteristics that could help explain the evolution of human behavior. (Remember . . . behavior doesn't fossilize.) Now this approach seemed chimeric. If we couldn't explain primate behavioral variations within the same species, how could we suggest, let alone claim, that early humans behaved like modern nonhuman primates? This stalemate made many scientists abandon primate studies.

Sociobiology, which integrated biology, behavior, genetics, and evolutionary theory, couldn't have come at a better time.[12] Sociobiology offered a "new" evolutionary scaffold, including genetics, that allowed scientists to make predictions about what an animal *should* do. Its chief metric was the reproductive success of individuals competing for limited resources. Variations within species, including primate groups, could now be explained as something to be *expected*, based on ecology, phylogeny, and reproductive strategies. Even better, the framework generated evolutionary hypotheses to test.

During Washburn, DeVore, and Hall's time, evolutionary arguments centered on what was "good for the group" or "for the good of the species." By contrast, sociobiology argued that group selection was impossible, that it was the *individual* that mattered most. When I thought of raiding in terms of individual benefits (and costs) using sociobiology's prime metric of reproductive success, I had a satisfying explanation for raiding. Raiding gave an individual better food, and this resulted in better condition, which in turn begat better growth and reproductive potential. Voilà, raiding was an evolutionary strategy. Aside from their nutritional benefits, human foods yield a day's worth of energy with a few minutes' effort, compared to the hours required by normal foraging. Raiders used those saved hours to socialize and rest, conserving energy rather than spending it on the search for natural food and its sometimes-laborious preparation. Therefore, raiders grew faster and both males and females weighed more than nonraiders. I couldn't say what happened to raider males when they transferred to other groups. I could say that during these years of study, raider females reached sexual maturity earlier, had first babies sooner, and had subsequent babies closer together.

But if raiding was a good strategy and not aberrant, why didn't *all* adolescent and young adult males become raiders? And why did only a handful of elder sisters leave their home group to join their raider male friends? What about the other low-ranking females besides Beth and her daughter? Wouldn't it make sense for them to join a smaller group with fewer females above them? Why did Billy choose his brothers over his mother?

The concept of animal "mind" had yet to gain credence—far from it—but the baboons' varied responses to raiding were taking me there. Sociobiology wanted to ground behavior in genetics and reproductive success, but this approach felt overly deterministic to me. Baboons had options, and how they picked one over the other was one of their most tantalizing mysteries. One thing I knew. Baboons are complex creatures, individually and, even more, in their relations with each other.

Over the past nine years, I'd watched baboons compete and defend using social strategies instead of aggression. This alerted me to their social complexity. The development of raiding added a new dimension to their behavioral repertoire. Now I could see the interplay between the *social* and the *ecological*. Here was evidence that baboons were weighing the risks and benefits of one action over another, not just socially, but as they decided what to eat and where to go. Yet mainstream animal studies reduced baboons to well-honed adaptive "machines," incapable of any sort of thought process. But if genetic programming explained all behavior, why didn't all baboons become raiders? Twenty-five years later, I was to revisit that question when I had to consider whether baboons faced with choices could also make mistakes (see chapter 12).

These were heady matters to ponder, but here on the ground Rome was burning. Raiding, already out of hand, was getting worse. I needed to find ways to stop the baboons from raiding and try to make peace with the farmers who now hated them. I tried not to think about failure.

Chapter 5

# The People Problem (1981–1984)

**AS THE RAIDERS GREW BOLDER**, tensions rose and, as I'd feared, conflict began. When farmers caught baboons "in the act," they retaliated. The angriest declared war on baboons. Maintaining scientific distance, I watched helplessly as farmers took up pangas* and went looking for baboons. The humans hunting baboons didn't discriminate between raiders and non-raiders. Many innocent animals were attacked, injured, and in some cases killed. Even now, I feel the sense of loss. These were baboons minding their own business, with no sense of impending peril. Individuals I'd watched for years were dying: Paul, Peggy's adolescent son; Big Sam, who'd learned the hard way about agonistic buffering; shy Naomi, Robin's mother; too many others. Settlements had sprung up near favorite baboon sleeping sites, making matters worse. I

*African machetes.

was always on edge, wondering when baboons would raid again, who would join in, how many farms they would hit, and if baboons would be hurt or die. This was my age of anxiety.

The raiding study had also promised to explore ways to prevent or stop baboon raiding, or at least manage it. With raiding now a full-fledged problem, this phase of the study became critically important. Did I find answers? I found some, and not all were the ones I'd hoped for. But I also saw that baboons, given choices, seemed capable of deliberation in a way I had not seen before this trying time. This was a silver lining, and there were a few others.

The control techniques we tested included traditional methods such as chasing monkeys from the fields, throwing stones, and sending dogs after them. These were modestly successful but required constant human vigilance. I also wanted to try a few newfangled devices. We tested oversized firecrackers called thunderflashers, and a system that loudly broadcast baboon alarm sounds, with middling results. I went so far as to lay dung from the baboons' arch-foe, the leopard, around the shambas. Surely this stuff would keep them away. It did not. At best, our deterrence methods, no matter how ingenious, confused the baboons. A colleague suggested "taste-aversion conditioning," which had worked well in North America to keep wolves from preying on sheep.[1] Accordingly, we laced selected plants with lithium chloride, an emetic salt that causes vomiting if consumed in large amounts. (We took care not to dose anything people would be eating.) It worked, sort of, but it didn't take long for baboons to identify the salty taste and avoid the doctored plants.

Raiding seemed a good foraging strategy from the baboon point of view, but it wasn't inevitable. Eburru Cliffs was a large group that frequently and mostly harmoniously interacted with Pumphouse.*

*Baboons have home ranges with porous boundaries, not defended territories like some primates and birds.

Many immigrant males came from Eburru Cliffs. The two troops' home ranges overlapped in the farming area.[2] As I watched Pumphouse and Wabaya activities around the farms, I could see that Eburru Cliffs was also interested. Or at least the large adolescent and young adult males were, and I saw them pull the rest of the troop toward the farms. That they succeeded was their misfortune. Farmer vengeance meant many Eburru Cliff animals lost their lives. But they learned. Eburru Cliffs abandoned the farming area and relocated to the farthest reaches of their home range, as far as possible from humans. I took note. Here was a solution to baboon-farmer conflict. Just run away.

By the end of the raiding study, we had clarity about at least one issue. For baboons, raiding was all about cost versus benefit. Baboons would continue to raid if the benefits reaped exceeded the costs paid. That meant raiding could be controlled if the pain of raiding outweighed its nutritional benefits. As we already knew, baboons were adept at evading costs, usually by outsmarting all of us. A guarded field turned buffet when the sentry went for a break or cup of tea. Even a brief visit to the hut unleashed raiders, patiently lurking nearby for hours and able to get a full day's calories in just a few minutes. On one occasion, raider males fooled human "chasers" charged with preventing the group from ascending the cliff to the planted fields. They fooled me too. The raiders had been below the sleeping site with the rest of the group, as if planning to hang out with the troop that day. Later, as we tracked the baboons, we noticed they were missing. We found them in the fields. They'd evidently peeled away from the group and circled around in the bush, out of sight, and made their way to a farm.

While the evolutionary benefits of raiding seemed clear, equally clear were the costs: primarily injuries, but also deaths, eleven in Pumphouse. As usual, baboons defied expectations because they did not act in lockstep. Far from it. Habitual raiders figured out how to avoid humans, thereby minimizing injury and death in their ranks. By adjusting so adroitly to the risks of raiding, the raiders were able to reap benefits that outweighed risks. Those many non- and occasional raiders found by angry farmers weren't so lucky.

I never killed a baboon as a deterrence method. It was something I could never do. But killing happened and raiding persisted. I realized that death doesn't count as a cost unless it involves two elements: The group must see the death and the death must clearly be linked to raiding. Otherwise, only the dead baboon learns a lesson and the troop might assume a missing male has simply set off for another troop.

I could now explain why baboons raided and offer suggestions about deterrence. Yet raiding seemed hopelessly entrenched and baboons were still dying. I turned to Jonah Western, who knew about these things. You may recall that I'd met Jonah, a conservationist, in the mid-1970s, when we were speakers at a seminar at the University of Nairobi. As it happened, we were both seeing others at the time, though there had been a spark.

The 1970s saw an influx of foreign researchers in Kenya, many young and working in wildlife research. A sort of culture evolved, centered on Amboseli, where Jonah's work was based. Every so often, gatherings happened there. I'll be honest. We blew off steam. There were picnics, game drives when we would pile into vehicles and motor out to watch local wildlife, and some pretty rollicking parties at night. Then I decided to organize a Halloween party at Kekopey. Jonah flew in, landing his Cessna at Kekopey's small airstrip. I'd thought maybe this would be our moment, but a foreign graduate student (dressed enticingly as a secretary bird) managed to distract him . . . for several years. Then, in 1979, I went to visit a friend at an old ranch outside Nairobi, near Jonah's place. We met up, and finally it was "our time." This was also around the time settlements started springing up on Kekopey.

Communications then were, of course, not what they are today. Jonah and I had trouble coordinating our schedules and hated the high phone bills. At one point, he showed up in California, thinking I was still teaching, but in fact I'd returned to Kenya. The only solution was to throw in our lot together, albeit as commuters. His house was in the outskirts of Nairobi, but the baboons still got most of my time. Sometimes he ferried me back and forth in the single-engine Cessna, flying through Hell's Gate near Lake Naivasha, with the wing tips nearly touching the walls of the canyon. It was as thrilling as it sounds.

Our friends didn't give us much of a chance. I was an extrovert, he was an introvert, but we shared an inclination to question orthodoxy about the things we studied. Ironically, these friends are all divorced or still single while we're still together after 40-plus years.

Jonah took on the conservation community when he studied the ecology of Amboseli—yet to be a national park—as well as the human ecology of Maasai pastoralists. Most conservationists were blaming the Maasai for the destruction of Amboseli's woodlands. Jonah's research challenged that view. He found that a rising water table was increasing soil salinity—a problem particularly for fever trees—and doing most of the damage, with elephants close behind. He took a lot of flak for supporting the pastoralists, but his science was sound and I believed him.

Our deferred courtship turned out to be serendipitous. I had a people problem and that was Jonah's specialty. Jonah's approach to conservation centered on working with people. He called it community-based conservation and he had a wise credo: "If people are part of the problem, then people have to be part of the solution" (see chapter 20).[3] Of course, I'd already had (mostly frustrated) dealings with farmers, but now I would take Jonah's path. I would alter my behavior in the hopes of altering theirs, looking to tamp down conflict with the baboons and, with luck, find solutions.

I scheduled meetings with individual farmers and explained the research results and suggested new ways of protecting their crops.

I listened to them and they listened to me. The results surprised and heartened me. Adversarial feelings—on both sides—almost magically vanished. In their place grew mutual respect and a collaborative sense that we were all in this together. All it took was the simple act of approaching people with the sincere intent to learn and help.

I looked into other ways farm folk could earn money so they'd be less reliant on crop income. The search led me to an Australian woman who came with her 10-year-old son to teach farmers to spin and dye wool and to weave rugs. We started a project called Woolcraft, specializing in rugs woven from wool colored by natural dyes from plants at Kekopey. The first rugs were rudimentary, but they got better over time. Though they no longer make rugs, farmers at Kekopey remain in the wool business, selling a naturally dyed spun product.

I could see that Jonah's community-centered approach was working, and it was working well. Dare I say this "social" alternative to conflict at least slightly resonated with the "making nice" behaviors I had seen baboons use to avert aggression? I helped the farm families build a primary school for their children, and I began to hire local Kenyans as research assistants instead of relying only on graduate students. Today this is a widespread approach, but it was novel in 1981. My new Kenyan research assistants did both crop raiding research and community relations with a sense of personal commitment that was lacking in the graduate students they'd replaced.

Raiding changed everything. It forced me to swerve away from purely academic questions and face up to "real life" embroilments with serious consequences. At stake were baboons' lives and farmers' livelihoods, existential situations for both.

I knew the tide had turned for the better when, at a meeting of farmers, William Murai stood up to speak. I feared the worst, since this man had been the most outspoken and bitterly hostile member of the farming community. But what he said, speaking for all, was that they would rather have the baboons and the Baboon Project than no baboons and no Baboon Project—because the project had brought so many benefits to the community. It was a great moment for me.

Crop raiding taught me two important lessons: Community-based conservation works to reduce conflict, and conservation is never "finished." Some efforts succeed, others fail, but there are few lasting solutions in this changing world and no end to new challenges. From here on, I would add conservation to my academic research.

---

Toward the end of my raiding study, after four years together, Jonah and I decided to marry. It was 1983 and almost spur of the moment. Jonah had asked several times, and then I did, and then it happened. Marriage was never something I'd yearned for, yet somehow, to my surprise, it felt right. The ceremony was small and simple, held in the modest but beautifully situated home of Jonah's friend, Philip Leakey. Yes, he was one of Louis Leakey's three sons, but unlike his famous brother, Richard, he had pursued politics instead of fossils. His place overlooked the gorge of the Mbagathi River that forms the south border of Nairobi National Park. Later, Jonah and I bought land there and built a house for ourselves.

We had chosen the end of February, during the long hot dry season, before the rains. But not that year. Rain came down in torrents and our borrowed four-wheel-drive car got stuck in the mud on the steep hill going to Philip's house. Under normal circumstances I would have jumped out and helped push. This time I flatly refused. I was not going to get married in a mud-spattered outfit, even though it was simple, hardly a big, billowy affair. The rest of the party heaved and we made it the final way to the house. The Kenyan marriage officer read the official words, including a stern warning not to commit bigamy, and Philip spoke briefly. I'd tracked down the last bottle of champagne in Nairobi for a toast, even though I knew Jonah, a nondrinker, wouldn't partake. The rest of our small group enjoyed the indulgence.

The much larger reception followed at the Carnivore, a Nairobi restaurant famous for its skewers of meat, including game meat, and popular with tourists. In retrospect, I appreciate how our guests represented an ideal mingling of every type of Kenyan, the outcome you'd want from work in the community. Jonah's Maasai friends were there, and my Kenyan field assistants, and many Nairobi friends, along with white colleagues and random expats. Back then, this sort of mix was atypical and somewhat scandalous. After food (sumptuous and tasty) and cake (dry and not so tasty), it was time to hit the dance floor. Not quite a bacchanal, but close. Jonah and I shared winks as we watched a colonial English friend, no spring chicken, dance the night away with a graciously accommodating Maasai warrior. Video cameras were yet to be the norm, so I only have still photos of the evening, but I smile each time I see them. Wedding gifts included a goat and a rooster that we later discreetly regifted to Maasai friends. But I still use the pressure cooker and mixing bowls we received as gifts more than 40 years ago.

---

Jonah and the wedding provided welcome breaks from the endless anxiety and pressures in the field. I realized I'd come full circle. I had opted not to study people as an anthropologist, but people, like it or not, were very much in the picture. And I needed their support. That prompted me to reflect on why I'd initially felt that the farmers, collectively, were my enemies. Years later, two stories in a Kenyan newspaper, the *Daily Nation*, gave me a new perspective. Above one, the headline screamed: "Human Misery as Drought, Hunger Ravage Counties."[4] Millions of Kenyans were facing starvation because the short rains were predicted to fail. The search for water and grazing would set off conflict as people, livestock, and wildlife competed for dwindling resources. Schools were closing and the devastating impact of the drought would be felt by countless families, even after rains finally returned. I wondered if the cycle would ever end.

I turned a few pages. "Starving Boy Died after Dog Meat Meal."[5] Puzzling headline. Did it mean that the dog meat was tainted? I read on, learning the tragic story of a single drought victim, young Mutinda Nzau. In a remote, famine-stricken village, 15-year-old Mutinda had not eaten a proper meal in five days. Desperate, he'd exhumed the body of his "best friend," his dog who had starved to death a few days earlier. He became ill soon after eating spoiled meat from the dog's corpse and died before help could be found.

I choked up. As the reporter warned, this story was a tearjerker. My tears were an outward sign of my social emotions—grief and compassion, sympathy and empathy. The images lingered all day and all night, invading my dreams. I awakened the next day with a feeling of dread.

Then I thought about the two stories and why my reactions were so different. After all, both were about human suffering. Why was I was emotionally stirred by the account of Mutinda and somewhat dispassionate about drought and starvation on a mass scale? Certainly, the suffering of many tens of thousands was a far greater tragedy than the death of a single boy. Yet it didn't feel that way. I had absorbed the drought story rationally, felt concern and donated to a charity helping the drought afflicted. But Mutinda's story had felt like a lived experience.

The difference lay in the presentations and their divergent appeals to emotion. While the first story conveyed distressing information about a catastrophic drought, it was essentially impersonal, devoid of names and faces. Mutinda's was different. His plight was described in poignant detail and included the lamentations of his mother, grandmother, and blind great-grandmother. I knew where he lived and how the family's meager supply of maize had run out days earlier. I felt his desperation as he made a meal of "man's best friend." I could identify.

These two stories, and my reactions, helped me understand why my attitude toward the Kekopey farmers had shifted toward understanding and acceptance. By the time farms arrived on Kekopey, I knew each baboon by name and shared a history with them. I was connected both scientifically and emotionally. But as the farmers filtered in, they remained essentially nameless and faceless to me, even though I had cursory dealings with some. I hated them because they killed baboons. The fact that I *knew* one group and not the other explained why I identified and empathized with one and not the other. I understood the Kekopey farmers' situation intellectually—yes, baboons stole their crops and livelihoods—but as long as farmers remained generic baboon foes, I could not feel for them.

Recent research confirms that identification and empathy work this way.[6] We often champion animals and children because they are less able to defend themselves. As humans, we have a strong response to injustice and unfairness.[7] We frame these violations using rudimentary concepts—good versus evil or perpetrators versus victims—and we look for something or someone to blame. Like the farmers, or the baboons. The evolutionary emotions that grease the wheels of social concern depend on these emotional responses.

Once I had met with and talked to the farmers—and saw through their eyes what the baboons had done to their fields—their struggles became real to me. I was meeting individuals where they lived, people with families, histories, and hopes. I could identify with them, feel their losses and their kindness, and be touched by their hospitality and generosity. An experience I'll never forget involved a visit to the mud brick hut of a farm family. They'd been hit hard by a baboon raid that day, yet greeted me warmly. They offered tea and laughed that I didn't take milk in it. They had roasted their very last maize cob and insisted I eat it. It gave them pleasure that I enjoyed it, and I made sure they knew I did. We walked their field, up and down the rows of tattered maize plants, and I listened as they told me how the baboons had eaten the few ripe cobs and pulled apart the immature ones, ruining the crop meant to sustain the family and provide seed for the next year. How could I not sympathize and feel terrible about what the baboons had done?

Now I saw both sides. I asked myself, Who actually has the right to this land? Jonah urged me to think outside the box, to get away from dichotomies where one side wins and the other loses. Coexistence was the only answer, but figuring out how people and baboons could live peacefully alongside each other would not be easy, not then and not now.

In a strange twist, my sympathy for the farmers now put me at odds with the growing animal rights movement. Ironically, to gain support, advocates for animal rights and welfare relied on scientific studies of animals in the wild. The early studies had kept it impersonal, labeling animal individuals and cohorts by age and sex, no names involved. But later and longer studies, particularly of primates, identified and named individual animals. A seismic shift occurred. People reading stories or seeing films about wildlife could connect and identify with animal individuals, especially named ones, and they cared. It had even happened to me, the scientist.

This scientific turn would have had little impact had the media not latched onto stories of individual wild animals, and they did because people responded. An international conservation movement formed around "charismatic" animals, notably apes, elephants, dolphins, and whales. It was hard *not* to care about Flo and Figan, Leakey and Wurzel (Goodall chimps); Digit and Titus (Fossey gorillas); Echo and Ely (elephants filmed at Amboseli National Park). Naming animals in the wild—cute names, at that—made their hardships real, their innocence compelling, and their rights and welfare worth fighting for.

But there was big wrinkle. Concern for animals was growing at the same time people were claiming habitats and fueling conflict with wildlife. Like me at first, the viewing public had little compassion for people who'd lost crops and livelihoods to baboon or elephant "pests." When an elephant was killed by poachers or in retaliation for crop damage, the world seemed to mourn. When an elephant trampled a person to death, only the family and community knew of it and grieved. A peculiar upside-down symmetry pushed by the media became a narrative widely embraced by the public: *Humanized* animals were confronting *dehumanized* people in a battle over resources. Who would have thought the simple act of naming a chimp or a baboon could create one of modern conservation's great dilemmas?

The questions "What is nature?" and "What is natural?" continued to vex me, and I consider them deeply in chapter 19. But in the early 1980s, I didn't dwell on them. I had a quiet laugh, though. The crop raiding study showed that these raider baboons—deemed "unnatural" by colleagues—were smart evolutionary actors, even if their new context was "unnatural." In my mind, the issue was semantic. "Aberrant" raiding behavior was simply a clever foraging strategy, albeit one that made them into enemies and put baboons in a new world of risk.

---

The sale and populating of Kekopey forever changed the baboons' known world and their lives. While some thrived, others needlessly died. It also launched my conservation career by giving me a window into the Anthropocene, the Human Age, transformation of habitats and landscapes. There were unexpected scientific benefits. Baboon individuals and troops reacted differently to the new resource, farms, and showed me new colors of baboon behavior. While evolutionary principles helped me make sense of some aspects of baboon raiding, the animals violated expectations based on existing evolutionary models and exposed cracks in those models. For a second time, I documented baboon options, this time in the ecological realm, and I saw them make choices. My interpretation of why they made those choices was essentially informed speculation. But there was nothing speculative about my conviction that baboon life contained complexities that deserved deeper and ongoing study. Some of these thoughts never would have occurred to me if humans had stayed where they were and not come into the baboons' world. Now I know their presence was all but inevitable.

From my perspective, the choice—to be or not to be a raider—seemed grander and more existential than consort tactics or agonistic buffering. The baboons had important decisions to make. *What kind of baboon will I be? Who will join me, and what will we do together?*

As my understanding of baboons grew, their responses to things they'd never seen before—to people suddenly farming where once they'd foraged—challenged me to think differently about evolution. Baboons made sophisticated choices about basics, like what to eat and whom to groom, and they proved flexible as conditions changed. They tested alternatives, negotiated with each other, and, seemingly guided by internal balance scales, weighed one possibility against another. While I lacked access to baboon thoughts, their negotiations and testing were visible to me "in the world," especially in this new world.

Though our outreach efforts vastly improved relations with the farmer community, the baboons weren't going anywhere; too many of them still wanted field crops and kept raiding no matter what we tried. I was faced with a decision that might have ended my credibility as a scientist, but what was happening on Kekopey could not be fixed, despite our good intentions.

Chapter 6

# Rehomed (1983–1984)

**JONAH, AN EXPERIENCED BUSH PILOT,** was much better than I at spotting clusters of animals, even an individual leopard or eagle. We were flying low and noisily in his single-prop plane over wildly varied Kenyan terrain. His had a talent called "search image," a very different skill from discerning distant baboons on the ground, at which I excelled. It was 1983, and I was on the hunt for a new home range for three baboon troops—132 animals in all—both miscreants and well behaved. I was happy when Jonah pointed out a troop of baboons trekking across a gully. Where other baboons lived, away from human habitation, would be a good bet for the Kekopey animals. I would need to track down the owner of this land.

How had it come to this? The Bad Guys, the crop-raiding Pumphouse splinter group, were now spending time at an army camp at the edge of Kekopey. Bored soldiers' wives initially welcomed baboon visits and enjoyed their hijinks. But before long, the baboons overplayed their hand, helping themselves to kitchen gardens and sneaking into married officers' kitchens. Amusement became irritation, then anger. Soldiers had already shot and killed several baboons. By this time, I was considered responsible for all baboon misdeeds. Word came down that army personnel had proposed a "final solution" for the baboon problem, a phrase I found particularly chilling. It was on me to prevent the slaughter of baboons who didn't learn to stay away. Yet another people problem; it was always a people problem, but the baboons would suffer most.[1]

Adding to our woes, farm raids continued despite our creative efforts to stop them. At least now we knew why certain methods did not work. Between ongoing crop raids and the new army problem, the situation had gone from barely manageable to outright dire. I had two options. One was the way of the "good" scientist: I could walk away, find new animals to study far from here, and leave these baboons to their fate. In which case army personnel would keep killing them. The baboons would eventually decide the wonderful food at the base wasn't worth dying for, but many more would die in the meantime. We had been through so much together, the baboons and I, and I'd learned so much from them. This option felt like abandonment, not good science.

The second option: I could physically move the study groups to another place, as far as possible from the temptations and harms of human habitation. It would be something like a wild, uncontrolled experiment.

Moving them would in essence be "playing God," but not an omniscient god. I had no idea whether such a move would have a good or bad outcome since there were no precedents to draw from. Endless internal debate. After 10 years of study, did I really know baboons well enough to pull off such a move, on this scale? I'd come to a good understanding of the baboon social world, but I had no way of knowing if their social order would hold or shatter in a different ecology with unfamiliar resources. Would they even survive? If I found an ideal location, then simply getting them there would be a herculean task. It's not like you can herd baboons onto a bus and whisk them away. I would need to find cages and experts to help capture them and arrange for their transport over rural Kenya's notoriously bad roads. Once released in their new place, wherever it was, I would somehow have to "convince" them to stay put. No wonder no one had ever done this.[2]

I put myself in the baboons' shoes, or maybe I imagined the baboons in mine. Given the choice, would I move to a foreign country to escape danger where I was, but with no assurance that the new, scarily unfamiliar place would be any safer? Or would I stay in a place that was comfortably known yet rife with risk? Baboon deaths could happen in either scenario, deaths I would feel responsible for. Alone with these deliberations, I lost sleep and sprouted a few gray hairs. I had troubling thoughts of my parents and their families in Eastern Europe, and the awful dilemmas they faced, because there was an inescapable parallel.

At last, the way forward took shape. As much as I resist anthropomorphism in science, my decision projected human values onto baboons, and not just any human values, but mine. I would take my chances in a new place. For the baboons, even a less hospitable natural setting would be better than Kekopey, as long as people and their enticements were far away. It would be an unprecedented baboon odyssey, if it worked. Admittedly, I was intrigued by the idea of studying the outcome of a move like this.

Even with decisions made, anxiety about the execution of the plan gnawed at me. Others—particularly Jonah—worried about me as I fretted about the baboons. On top of it all, I was pregnant for the first time, in my late thirties. We had scheduled travel to California for our daughter's birth so she would have dual citizenship and access to advanced medical care. It was almost too much.

But a part of me, the scientist, felt a building excitement. I knew that tracking these familiar baboons as they adjusted to a new home would present an extraordinary research opportunity. Which skills, traits, and habits would prove the most helpful to them as they made new lives for themselves in a very different ecosystem—and would they be enough? Most important to me, would they survive? I'd know soon enough.

---

In conservation parlance, "translocate" means to move animals from one place in the wild to another place in the wild, to an area where the species has a history. At the time, translocation was a new conservation technique, devised to stem the tide of species loss. Today, an entire literature is devoted to translocation as a conservation and management tool, but in 1984 there were no precedents for what I was attempting, no blueprint to follow. The few on record weren't comparable: several haphazard translocations of orphaned and injured animals, small bird populations, a rhino here and there, some successful, some not. I hoped to save—that is, conserve—the baboons, but I also wanted to study them over time in their new setting, to track their response to this major disruption in their lives. My effort would write a new chapter.

Baboons have an important attribute that factored into the decision to leave, namely their "generalist" approach to feeding. A wide array of foods can sustain them, and they aren't terribly picky, despite having definite preferences. This made them adaptable and greatly improved their odds for survival. But if a species as versatile as baboons failed in a new place, it would call into question translocation as a tool to save any threatened or endangered primate group, especially species with specialized dietary needs. On the other hand, if the baboons survived, this translocation could serve as a model. I'd figure out the dos and don'ts of an operation like this, in effect create the handbook. Other translocations could tailor the process to fit their species' needs.

But I get ahead of myself. I hadn't yet found a new place for the baboons, and it wasn't going to be easy. After many white-knuckle flights in small planes, scanning ranches and meeting with ranchers, I'd learned something: While I was falling in love with baboons, most people were wishing they didn't exist. I could understand farmers wanting nothing to do with a baboon project. But I didn't expect stiff resistance from large ranch owners and managers on the Laikipia Plateau, people who controlled tens of thousands of mostly natural acres, even after I assured them that baboons and livestock peacefully coexisted on Kekopey. Most were a good deal less pleasant than the typical baboon, despite my best persuasive efforts. At Kekopey, I had found myself bearing the brunt of ill will toward baboons before working with the community. From that experience, I'd learned not to take anti-baboon sentiment personally. But the search for their new home brought me in contact with ranchers who pushed me to my limit.

One such person was Mr. H., who ran a large ranch that brimmed with wildlife, baboon foods, and potential sleeping sites. Initially pleasant, he turned apoplectic at the first mention of baboons. I was just about to shoot those bloody baboons, he raged. His tirade included "relevant" facts. Did I know that when baboons pee in water troughs cattle won't drink the water?* Baboons also tear up grassland, he sputtered, clearly unaware that baboons *help* by digging for sedge corms that compete with grasses favored by cattle. I hardly knew whether to laugh or cry and removed myself as quickly as possible. Some owners thought I was offering to *remove* their baboons, and were thrilled at the prospect. I wrote a mental memo to myself. Baboons needed reputational rehabilitation. I would make that part of my agenda.

Finally, I found a rancher with an open mind. A small and energetic man, Mr. J. had a large ranch in Mukogodo, on the eastern

*Baboons rarely pee in water troughs, but it is a small addition to a large volume of water that doesn't dissuade other baboons or cattle from drinking.

Laikipia Plateau, about 125 miles from Kekopey. He'd had a baboon as a pet and actually liked them. He listened intently to my pitch and had smart questions and comments. Yes, the baboons would cost his ranching operation nothing, while the Baboon Project would bring benefits, including watchful eyes in remote parts of his sprawling (and spectacular) ranch. Besides expansive grasslands, this part of Mukogodo had many more tree species than Kekopey, and granite outcrops called kopjes where the baboons could sleep. There were even a few little dams with water, so they would not go thirsty, even in the dry season. It almost seemed too good to be true. But it was true, and my search had come to an end. The baboons' new home would be on the ranch called Chololo.

And so, in August and September of 1984, I translocated three baboon troops from Kekopey to the Laikipia Plateau in north-central Kenya. There was a total of 132 animals: Pumphouse, Bad Guys, and a third troop I called Cripple Troop because so many of its animals had been maimed by high-tension power lines recently strung across an edge of Kekopey. Cripple Troop members were the first to visit the army camp, but now raiding at the army had swept like a contagion through the local baboon population. The addition of Cripple as a study group would serve as my translocation "proof of concept." I'd learn by capturing and moving them first.[3]

---

For decades, professional trappers had captured baboons for biomedical research. Despite their expertise and special equipment, these pros never managed to collect more than half of any group. I needed to nab every member of three troops. I applied my knowledge of baboons to devise ways to improve the capture rate. First, the baboons needed to be convinced the traps weren't a danger. Standard traps were low-tech affairs made of wire mesh with a small wooden platform attached high up on an inner wall. To lure the baboon inside, a maize cob would be attached to a string threaded through a hole in the wooden platform. The other end of the string was attached to the open door. A baboon had to be fully inside the upright trap to reach the cob. When it grabbed the maize, the string broke and the door dropped—simple but effective.

I ranked the three troops according to their importance to my research. Cripple Troop, studied for the shortest time, would be the guinea pigs, the first to be trapped and transported. The Bad Guys would be next. Finally, I would tackle Pumphouse, the largest of the troops and the one I considered most scientifically valuable.

The baboons would be naturally wary of the traps. I adopted a strategy long used by professional trappers: baiting prior to capture. The bait had to be sufficiently "attractive" to overcome the baboons' natural caution. Maize kernels, like caviar for baboons, were the logical choice. We sprinkled them on the traps' floors and platforms and watched as the baboons moved freely in and out of the traps to claim them. The baiting weeks continued beyond the several days recommended by the pros, anything to boost the odds for success. Professional trappers told us that you only get one shot; any baboons that elude capture will warn others away from the traps.

I thought we needed as many traps as animals. That was fine for the two smaller troops, but Pumphouse was as large as Bad Guys and Cripple combined. I counted again, but we were coming up short on traps, on loan from the Institute of Primate Research in Nairobi (IPR).* Then I realized that if mothers came into the traps with their babies, which was likely, we wouldn't need a cage for each. But I was still short a few traps for Pumphouse—and beginning to wonder how logistics had become such a big part of my job.

**August 1984**: I perched myself on a lookout rock and watched from afar toward Cripple Troop's sleeping site. I didn't want to be seen or associated with any trauma that capture and translocation

*The organization is now known as the Kenya Institute of Primate Research. For more information, see https://primateresearch.org/.

might cause, so I let the team from the IPR do the trapping. It was getting late and the baboons should already have left their sleeping cliffs. Had I misjudged? Then there they were, trotting into the clearing. This time it wasn't maize kernels in the traps but irresistible whole cobs. One by one, the baboons slipped into the cages, snatched the cobs, and triggered the doors. I worried that the sound of the doors snapping shut might spook them, but they merely blinked, then went back to their munching, oblivious to their detention.

Once all of Cripple Troop had been trapped, the IPR team went to work. Quietly and deftly, team members pricked each animal, except babies, with a syringe of sedative fastened to the end of a long pole. The animals recoiled when approached, but quickly lost consciousness. So far, things were going as planned. Unsedated babies were strapped to their mothers with string and clung on for dear life. The sedated group was then loaded into a pickup truck bound for a holding area. There the animals were transferred into smaller cages. Before recaging, each animal, while still asleep, was weighed and examined by a vet.* Blood samples were taken for later analysis. The baboons were then ready for convoy transport to their new home. Things had gone surprisingly smoothly.

After trapping and transporting Cripple Troop and Bad Guys, I felt expertly prepared for Pumphouse. I'd also come up with an idea for getting around the trap shortage: I would trap several males away from the rest of the troop and process them before the others. I set up three traps near the research house, knowing that large males would often swing by, hoping for an open door or window and a shot at the human food inside. I'd had firsthand experience. I was working at my desk one day and suddenly felt a presence in the room. I turned to see Sterling, a recent transfer from Cripple

*This was when I learned that the raiders weighed more and were in better condition than the non-raiders.

Troop, reaching for a bunch of bananas on the dining table. I looked at him in astonishment. He stared back at me, almost daring me to do something. By the time I leaped from the chair to shoo him out, he was already scooting through the open door with several bananas in his mouth and extras tucked in his armpits.

So, we positioned the traps close to the house and set them when I knew Pumphouse was foraging nearby. As if on cue, the males appeared. Each aimed for a different trap and the maize cobs within. The doors fell and they were caught. The IPR team took them to a holding area some distance away from other Pumphouse members, then sedated and transferred them to holding cages. Voilà, we had enough cages for the rest of Pumphouse.

Only one animal stayed behind, a very smart male named Arnold, easily recognized by his blond hair. Arnold had figured out how to get the maize cob and avoid entrapment: He reached for the cob with one hand and propped the door open with the other. We found him methodically moving from trap to trap, collecting a maize bounty. Improvisation was called for and I had an inspiration. The IPR team set up a trap with a trip wire rigged to be manually controlled by a person. When Arnold entered the trap and reached for the maize, we would drop the door. I was convinced this would work, but it did not, nor did any other trick we tried. In the end, he stayed behind. I was confident he'd easily find a new troop to join, and humbled at having been outsmarted, yet again, by a baboon.

---

The drive to the new place was an adventure in itself. Each of the three trips involved a convoy composed of a large truck carrying caged baboons, then me in an Isuzu Trooper, then a pickup from IPR. Only half the distance was on tarmac, or barely paved roads, where baboons had their first and only experience of potholes. The remaining 50 or so miles were rutted dirt roads that narrowed to a single lane as we neared our destination. September marks the end of the long, cold dry season, so conditions were mild, but I made sure the baboons were well watered. The cages were covered with tarps so they couldn't see where we were going. I didn't want the baboons trekking back to Kekopey.

The final challenge was no less difficult than trapping and transporting masses of baboons. I needed to make sure they didn't stray from their new home, because I had no agreements with adjacent landowners. Here again, I used my baboon knowledge to design the release. The goal was to minimize what might prompt them to run away. I relied on what I knew about the differences between male and female baboons—specifically that females as a group are cautious creatures. If I released females and young first, but kept the large males captive, I reasoned that the females would linger nearby, afraid to venture far without male protection.

My next move was more speculative. Since males transfer between groups, they're less attached than females to a troop or a home range. But if the females settled down near the caged males, they might act as a check on the male tendency to wander when set free. If that failed, I had no plan B. During this time, I broke a golden rule of my primate study, but only for a short while: I fed the baboons as a top-up as they took the time they needed to discover natural foods in this strange new place. We used feed meant for young livestock—nutritious but less delectable than maize—and broadcast them across a wide area to avoid provoking aggressive competition.

Within a month, I knew the strategy had worked. The baboons were settling into their new home without much fuss. Descendants of the translocated animals still live here, studied to this day. In retrospect, I can think of information we might have collected that would be useful now, but at the time concern about the baboons' survival was all-consuming. I was counting on baboon intelligence and adaptability to help them figure out where to sleep, what to eat, what to fear, and how to fit into a place that was already home to other baboons. However things turned out for them, I would have a record of the process and their progress. An entirely new area of study had opened up to me.

Still, I had prepared for the worst. We fitted each baboon with a small plastic ear tag, different colors for each troop. Each tag had a number to enable the identification of a baboon should it die after translocation. I would be able to identify a dead baboon only for a few days until the animal's distinguishing features—face, tail, hair color, scars—had degraded. Fortunately, the tags proved unnecessary.

Several adults in each troop also wore radio collars. This made it easier to locate baboon groups during the adjustment period, when troop movements were unpredictable. Their use required me to climb to a high point to catch the signal, then rush down a steep slope to reach the baboons before they disappeared on their daily walkabout. As the baboon troops settled in, radio tracking became less important. I could return to the old ways: finding them at sunrise and following them until they returned to a sleeping site.

There were amusing moments. I planned all along to release the baboons late in the day, below a sleeping site that was also within sight of water. I'd just sprung the Bad Guys, and as I predicted, they ran up the rocks. But as daylight dimmed, I didn't see the usual "settling in" prelude to slumber. Instead, the adults paced back and forth, put their bottoms on the rocks, rubbed side to side, and looked around, clearly disgruntled. The kopjes, granite outcrops, were seemingly less accommodating than the rocky cliffs at Kekopey where they had previously slept. I left them when it got too dark to see. They slept through the night on that kopje, but the next morning, everyone seemed tired. After a few days, they'd made peace with their new sleeping arrangement. What had been new and different was slowly becoming familiar.[4]

I too needed a new place to sleep during my days with the baboons. An old house that had belonged to Geoffrey, one of the two initial owners of Chololo, was my original base. It was a dilapidated building of mud brick and split logs with a steep roof covered with rusty old iron sheets. It looked to me like an Australian outback settler home. Though definitely better than a tent, it was rougher than the research house at Kekopey. I would make do.

And so began my new baboon study, which I now thought of and described as an experiment, and a bold one at that. I'd moved many baboons to a new ecosystem with no foreknowledge of the outcome, the very essence of an experiment. Now my objective was to learn how baboons would adapt to an ecology with unfamiliar resources. I would finally get answers to the questions that kept me awake all those many nights. As it turned out, there would be many more questions I hadn't thought to ask.

Chapter 7

# Strangers in a Strange Land (1984–1988)

**CONVINCED THE BABOONS** would survive at Chololo, I set about finding answers to the questions my new study posed: In what ways, if any, would translocation change their behavior? Would they still be "normal" baboons? No longer could they be considered "natural." Remember, colleagues told me that the baboons were "contaminated" by crop raiding and would be doubly so by this forcible move. Would the raiders revert to natural baboon behavior? Would the non-raiders behave differently? Most importantly, would the baboons' social structure remain intact, or would competition for vital yet unfamiliar resources tear it apart? The answer to this last question would force me to challenge orthodox thinking about a fundamental mechanism of evolution.

To begin this new investigation, I decided to use the same scans I'd developed when comparing raiders and non-raiders at Kekopey. A scan is like a snapshot of the whole group taken at a moment in time. To get each "picture," an observer quickly walks through the group for 10 minutes, noting on paper what each baboon is doing: grooming, feeding, resting, traveling, and so on. If feeding, what food is the animal eating? It can get complicated. The proper notation of "food" includes not only the plant species but also the part of the plant being consumed, such as root, fruit, leaf, flower bud, flower, or seed. These are important details. Information collected by strictly controlled methods like scans can be explored statistically later. From that analysis, a behavioral inventory and diet variations can be established and compared.

First, I contrasted activity "budgets" of raiders and non-raiders after translocation.[1] Significantly, all the raiders were in better condition. Data collected during the translocation showed that they weighed more than non-raiders and they were healthier. Activity scan data would tell me whether raiders and non-raiders were selecting different foods in their new home. I didn't think they would, but moving forward I could also compare different groups, different seasons, different years, and different individuals.

Of course, I couldn't do all this by myself. My team grew from a core of four research assistants at the crop-raiding project to six at the new location, headed up by Josiah, who had now been watching baboons for five years. These were local folk, most without a secondary school education, but they came from subsistence farmers who also had livestock so they knew individual animals. For them, learning to identify baboons came naturally, and they were already familiar with many of the plants. The training period took at least one year and included weekly meetings where we checked for consistency in data collection. I could return to my teaching post in California each spring, secure in the knowledge that the project was in good hands and that data collection would proceed uninterrupted.

I was also increasingly pregnant, more evident by the day. I had to ask the "guys" to give me a hand getting up the dry gullies and to please cut me some slack if hormones got the best of me. It was all more than a little astonishing. First a great baboon migration, now the heady prospect of motherhood. I was determined to be as attentive as the typical baboon mother. I also wanted to apply the lessons I'd learned watching them. I had concluded, for example, that pacifiers are fine; baboon babies spend much of their time on mom's nipple, not always feeding. It just made sense.

---

The Bad Guys needed a different name in their new home. My Swahili still wasn't very good because everyone wanted to practice their English with me. But my instincts were right. Why not "The Angel(s)," Malaika?* And angels they turned out to be, once crops were out of the picture. Pumphouse was also on its best behavior, but then they'd always been an exemplary group.

By day, Pumphouse and Malaika stuck close to each other, even though they'd been released near different sleeping sites and continued to sleep in different places. Normally, baboon troops tend to keep to themselves, but the move seemed to have them looking for security in the familiar. As they adjusted to their new surroundings, it made sense that the two translocated troops would shadow each other and feel safer that way.

Pumphouse and Malaika, sometimes alone and sometimes together, explored their new terrain and gradually expanded their range. This was a good sign. Each home range now covered more than 10 square miles, larger than on Kekopey. The new landscape was diverse. There were flat lands with deep soils as well as arid areas to the east marked by slopes, gullies, ravines, and dry sand rivers. This was the communal land known as the Ndorobo Reserve—later, Mukogodo—a region traversed seasonally by Maasai cattle-herding pastoralists. To my surprise, the thin soil

*Later I learned that it should have been plural, Walaika.

topography of the communal land supported a large variety of plants, including many adapted to arid conditions. The problem was that every plant here had thorns. I traded shorts for long pants.

The kopjes were magnificent and magical. A large stone might balance precariously on the edge of another rock. Why didn't it fall? Kopjes were originally rock encased in soil that slowly weathered away, revealing an array of sculptured shapes. Once exposed, slabs of rock could peel away like layers of onion. Rock outcrops dotted the landscape in every direction. Also visible were the remains of temporary Maasai settlements, *bomas*, which acted like nutrient hotspots in the rains. Finally, Mount Kenya was a glorious backdrop to replace Lake Elementeita and Lord Delamere's Nose at Kekopey. A colleague from Kekopey remarked when he visited, "You moved them so you could work in this beautiful place." He had a point.

But now, in October 1984, scenic beauty belied a serious food shortage for baboons and other wildlife. In a stunning stroke of bad luck, we had translocated during the worst drought in 20 years. Nuclear winter seemed to have descended on Mukogodo. Trees had dropped their leaves (not that baboons eat tree leaves—they're barely digestible for baboons), nor did they offer fruits, pods, flowers, seeds, or gum—basic baboon foods. The grasses were white, not green or golden, totally devoid of nutrients. Could the baboons survive, let alone thrive, in these harsh conditions? Maybe moving them *had* been a mistake. I continued to supplement the baboons' diet for a few weeks to ease the transition and help bond them to this new place. The supplement fell short of meeting their daily requirements, but it helped sustain them as they learned about local food sources. Finally, the short rains arrived and Mukogodo erupted into green.

How would the baboons create their new diet? I watched the process unfold. The baboons added new foods slowly. They concentrated first on what they already knew: *Cynodon* grasses are high in protein, particularly the new green blades and stem-bases. I had tasted the sweetness of fresh *Cynodon* on Kekopey. They also ate the flowers and green seedpods, and later the dried seeds from the pods of *Acacia xanthophloea,* the fever tree, which had been part of their Kekopey diet. Years before, I'd nibbled a few fever tree flowers. Despite their heavenly scent, the flowers only have a trace of sweetness and the unfortunate consistency of cotton wool. I understood why the troop went looking for water after eating hundreds of them.

The baboon diet was expanding from plant species they knew to unfamiliar plants that looked similar. Mukogodo had more acacia tree species than did Kekopey. I was slower than the baboons to learn to tell them apart. Each had distinctive seed pods, but pods stayed on the tree for only a short time. Leaves of the different species all looked the same to me. Eventually, I used trunk shape and color to sort the smaller acacias. Eating acacia foods illustrated the baboon "rule" of familiarity for expanding their diet. To my mind, given the baboon template of what an acacia should look like, thinking of wait-a-bit with its very different leaves and flowers as food was a big leap, but one they eventually took.

They found stands of prickly pear cactus, *Opuntia vulgaris,* same as on Kekopey. The baboons readily ate the moisture-laden cactus pads after biting off the thorns, but there weren't enough cactus stands to get them through the dry season. That's when I noticed that they had hit on another food-finding stratagem, this one involving indigenous males who'd begun to join the translocated groups. The translocated baboons closely watched the local males, who of course knew the terrain well, and they observed as the locals partook of a ubiquitous but unfamiliar succulent, *Sansevieria.* Sansevieria, with its fleshy edible part hidden underground, was strikingly different from the foods baboons knew at Kekopey. It took practice and strength to find a spike that could be pulled from the hard ground, but the reward looked like celery and, like celery, was full of water. The domesticated variety is commonly called snake plant or mother-in-law's tongue, perhaps because the long erect leaves end in a sharp spike. Slowly, other succulents, previously ignored by the baboons, became "food." And, yes, I was becoming an accomplished amateur botanist. Secure water sources were, of course, essential. I made sure dams were visible from each release site. But as the baboons ranged farther out, they often found themselves far from water. Again, indigenous males served as instructors. They knew to dig holes in dry sand rivers and tap into water just below the surface. Indigenous baboons hadn't been part of my translocation strategy, but they turned out to be vital to the translocation's success. Not only did they help the newcomers find food and water, they also provided unrelated males for the gene pool.

Pumphouse and Malaika often tagged along behind local troops. They kept a respectful distance but paid close attention. Soon, they were tracking the same foods through the landscape on their own, in a literal case of monkey see, monkey do. This serendipitous mentorship of local baboons no doubt saved translocated baboon lives. It also gave me insight into baboon learning.

Moving the baboons ensured their survival, but I benefited as well. I had the best seat in the house for viewing the drama of baboon adaptation to a new and very different ecology. Was it adaptability or adaptation? That's not an easy question to answer (see chapter 12). Most primate field research, particularly the sociobiological hypothesis-testing type, concentrates on the *outcomes* of behavior, such as who wins a sexually receptive female or who gets to feed in the acacia tree full of flowers. These findings are used to extrapolate individual fitness and predict reproductive success. I was increasingly convinced that *process* was important, not just outcomes. That meant observing the full array of baboon behaviors during this adjustment period—in real time, over time—as well as the outcomes of individual behaviors.

What most surprised—and impressed—me was that the baboons did not fight each other for food, which was scarcer here than at Kekopey.[2] Some would argue the translocation was too sudden and unnatural to be a valid test of natural selection. Perhaps. But even in the years following the translocation, competition for food was rare. I only saw hints of struggle when key foods were just appearing or just about to disappear, staples like acacia flowers or new green grass flushing on old bomas sites.

During the rains, a baboon—or baboon group—can easily find enough food with just a random walk. As the dry season progresses, food sources shrink and become more dispersed, making it impossible for any individual or group to monopolize what little is there. Recurrent droughts should have been the periods of highest competition for food, but they were not. Instead, baboons made a tactical shift away from food foraging to energy conservation. Social tensions melted away, replaced by lassitude. Males, typically vying for receptive females, rested quietly side by side, disregarding the rule about personal space. Females no longer competed for proximity to babies, and youngsters stopped playing. This was not "nature red in tooth and claw," nor was it "survival of the fittest," because there was little competition. This was how they endured a time of scarcity, together.

The second surprise was that the baboons "managed" the shock of the move by relying on each other, on their social group, on collaboration, and on knowledge acquired from local baboons. No one died of starvation, the expected outcome for weaker and lower-ranking animals in a classic evolutionary scenario. Even babies of both low-ranking and high-ranking females survived.

The translocation may have been unnatural, but it shined a light on the baboon skills and abilities that ensured their survival in a new place. No new skills appeared, nor were they necessary. Baboons solved novel problems in real time using the "tools" they brought with them; they didn't wait for natural selection over evolutionary time to reconfigure their genes to tell them what to do. Their individual adaptability and flexibility, along with their "interpersonal" resources, were key to their survival in these extremely challenging conditions.

---

In the years following the translocation, as I pondered its significance, my thoughts turned to an important evolutionary principle. What did "survival of the fittest" mean for a baboon?

The "baboon model" gained favor when genetics entered evolutionary studies. Charles Darwin and Alfred Russel Wallace had relied on evidence from geology and paleontology to arrive at the notion of evolutionary change.[3] Darwin traced the fossil record and examined dissections of embryos and older individuals. This provided additional evidence of change through time, namely "evolution" and "adaptation." As to behavior, evolutionary interpretations were only educated guesses. That's why my mentor Washburn was so excited about the baboon model. With rare precision, it portrayed the evolutionary meshing of anatomy, behavior, and ecology in a single, well-adapted animal, the baboon. This gave the model relevance to the study of human evolution.

In the mid-1970s, at the beginning of the "sociobiological era," individual animals were called "gene machines." Richard Dawkins famously claimed that a body was the gene's way of making other genes.[4] Genes created sophisticated behavior: "Free will" played no part in animal action. Earlier, Darwin borrowed from his contemporary, Thomas Malthus, the idea that resources were always limited, which meant individuals always competed.[5] In the new sociobiological model, "fitness" ordained the success, or not, of individuals as they competed for limited resources and the chance to get their genes into the next generation. At the time of the baboon translocation, this was the widely accepted version of natural selection.

But after situating the baboons in their new home, I didn't see selfish individuals competing for limited resources. I saw just the opposite. Was it the timescale that made a difference?

I remembered how Washburn challenged students to think about time differently. He offered different ways to conceptualize evolutionary time. For example, on a graph where all of geologic time is compressed into a year, our most ancient human ancestor appears at 8:00 p.m. on December 31. *Homo sapiens* shows up a mere 10 minutes before the year ends. I "got" evolutionary time, but the baboons convinced me that different timescales were relevant to different questions, that there was no "one size fits all" timeline when breaking down the components of evolution. "Baboon time" was unfolding as we watched, propelled by the translocation. The pressures of the translocation, the need to adapt quickly, were giving me a glimpse of evolution set on fast forward. What a distinct contrast to evolutionary time, which unfurls over tens of thousands or millions of years.

Clearly, an evolutionary timescale makes sense for geology and paleontology. Does it make sense for behavior? I thought through the logic of the argument. Behavior doesn't fossilize. Okay. We can gain insights about types of behaviors shared by all mammals (like a close mother-infant bond), or by all primates (the need to find safe sleeping sites), and a variety of other shared behavioral "outcomes" of evolution. Yes. But can we pinpoint in time behavioral inflection points, when shifts occurred in the ways animals related to each other or their surroundings? Clearly not. Even Washburn wasn't thinking on this scale when he relaunched primate field studies.[6] He was asking questions about BIG patterns—how an animal uses its body moving through space, why baboons live in big groups, what is the function of a male baboon's mantle of hair and large canines. I'd bought into Washburn's argument that baboons might be good models for the beginning of the "human experiment." Many scientists still think about chimpanzees or bonobos in this way. Today I am more cautious. Watching the baboons made me think not so much about "who" gets into the future, but "how." What does a creature like a baboon, living in real time, in a real troop, with a history of constraints and opportunities, have to do to *get into the future*, to become part of evolutionary calculations? Not just have babies, I thought, but also learn to behave in certain ways.

---

I couldn't stop thinking about the Cripple Troop reunion. Ordinarily, when baboons meet up after a brief separation, you'll hear a casual exchange of grunts. A more prolonged separation might end with lipsmacks and quick hugs as well. This was different. When I released Cripple Troop's males after the first translocation, a portion of the troop somehow scattered and vanished. I stayed with the remaining group, trying to concentrate on note-taking, but worried to distraction about the missing members. If they didn't make it back by nightfall, if they truly were lost, both groups risked predation. Then, near sundown, I looked up and saw the lost ones coming through the grass, as if by chance, but I knew they'd been searching for their troopmates. Seeing them, the group I was with froze. Then, like magnetized bits of iron, the two parts of Cripple Troop rushed to greet each other, not just their friends and family, but animals outside their cohorts. The rounds of embraces, lipsmacks, and grunts were unlike anything I'd ever witnessed before or since, so much like a joyous human homecoming. Separation in this strange new setting had clearly led to amplified expressions of relief, pleasure, and connectedness. Reassuringly, I saw that their social world was still intact, despite their lives being thrown asunder—and maybe it was stronger than before.

I had found another piece of Darwin's monkey puzzle. Recall that Darwin had observed an old male baboon risk his life to rescue a young baboon from pack of dogs, a behavior he admired but could not fathom.[7] It wasn't any old male, but a male friend of that young baboon. I'm convinced that Darwin saw, but did not recognize, *survival through the social*—not survival of the fittest, in the narrow "Darwinian" or sociobiological sense. The translocated baboons survived by collaborating. They relied on shared, portable social knowledge, supplemented with learning from indigenous baboons. The Cripple Troop reunion told me that baboons, not just humans, have social emotions that grow from mutual need and interdependence. In fact, it looked to me like feelings of connectedness had underpinned their survival. Put another way, the "social" was a *resource* that carried them through their adjustment, their terra firma when everything else had changed. The social group seemed to me an actual mechanism of baboon adaptation.

Because *the group* appeared so clearly to be the adaptive vehicle, I realized the benefit of moving entire groups rather than individuals, or artificially created groups, which moves have notorious failure rates.[8] Our natural groups of baboons succeeded beyond all expectations. They knew and trusted each other, and together they met the challenges of learning how to survive in a very different place.

I often joke that the baboons read the scientific literature and then do the opposite. The translocation was one of those times.

---

Crop raiding and the translocation introduced me to an aspect of myself I hadn't properly known. I ignored my colleagues' advice to drop these baboons and find a new "natural" group of primates to study. Doing so wasn't an option. I couldn't abandon these animals. I knew them well, they knew me, and together we had been through so much. I cared about them. Once resolved, I stopped thinking about the prospect of failure, though others didn't. Even Jonah wasn't convinced the effort would work, though he offered support throughout and only told me about his doubts afterward. It turned out that I had convictions stronger than my need for approval, even if that made me something of a scientific renegade.

I also realized that the "naming" of individual baboons had unintended consequences. A bond formed, and with it an emotional investment. The Little Prince understood.[9] The fox said to the Little Prince, "One only understands the things that one tames." The prince muses, "My fox when I first knew him was only a fox like a hundred thousand other foxes. But I have made him my friend and now he is unique in all the world." The fox then admonishes the Little Prince: "You become responsible, forever, for what you have tamed. Men have forgotten this truth . . . but you must not forget it."

Was this the reason I couldn't give up on the baboons? I had named them as part of a new type of scientific study. But I had also, in a sense, "tamed" them. By 1973 I had habituated the troop so I could walk with them. That forged a connection between us, and they showed me their true selves. Now the baboons were special to me. Fear and aggression weren't the only signals to cross the species barrier—as assumed at the time. I felt empathy, sympathy, and compassion toward them. Part of my scientific training told me that emotions had no place in "objective" scientific studies, yet I felt emotional about the baboons. Two *me's* now lived side by side, the scientific me and the emotional me.

Slowly, I came to accept that I could feel strongly about my subjects and still do rigorous science. I've often been asked how I could watch a baboon die and do nothing. It is always hard, this shifting between the two "me's," and learning to do it took time and practice. My early tape recordings of predation on young Thomson's gazelles objectively described what was going on—"Rad held the baby down to the ground. He bit into the soft belly with his incisors"—and also captured my sobs. Yes, the emotional me sided with the gazelle and not the baboon, even though I knew that in "science" you don't side with either. Only later did I realize that the two me's didn't cancel out each other. I could do good science and care about my subjects. Caring never stopped me from being thorough and impartial when I needed to be.

I had become part of the baboons' world in ways subtle and unsubtle. I had inured them to my presence. They went about their business as I walked among them taking notes. A few times I chased a farmer's dogs that were harassing the baboons near a little shamba. I did this without thinking, even though it violated the rules of scientific observation. I had even moved them to a new home. The "emotional" me sometimes took over. I tried very hard to think before I acted, and most of the time I did.

What were the consequences of my actions? The baboons trusted me—and those I brought out with me; I was a visitor's "passport." Did this make them relax their guard? Was I partly to blame for the deaths during crop raiding? Without a gun—only the army had guns—a farmer had to kill a baboon at close range with a handheld weapon. I worried that taming the baboons made it easier for that to happen, since they trusted this human, me. That is why the Little Prince was right: You are responsible for what you tame.

---

The success of the translocation let both "me's" relax. Secure in their new home, the baboons' future was theirs to create. We all settled in. I could enjoy watching and learning again, so I was totally unprepared for the next surprise, which involved baboons and elephants.

PART II

# Dispatches from the Field

Chapter 8

# Troop Movements (1986–2000)

**WITH THE BABOONS** settled in at Chololo, I shifted my focus to a baboon phenomenon I had wondered about and yearned to understand. My earlier studies had elucidated how the social meshed with the ecological. Now I wanted a better understanding of baboon decision-making, specifically how the group decides where to go each morning. It was clearly a complex process, involving intense negotiations that sometimes lasted hours. It seemed telling of so many things about baboons. Some baboon individuals stood out from the group and exerted influence, others kept trying and failed. Some simply bided their time as the discussion went on. In the end, no matter how fractious and disorganized things got, the group reached a decisive moment of unanimity. But what exactly were the dynamics at play? Why did some lead and others follow?

I also wanted to know how these daily movements figured into the still unstudied question of how a baboon troop establishes the perimeters of its range. Here the translocation offered another remarkable opportunity. Freshly arrived, the animals in the three study groups would need to create new boundaries in which to range, taking into consideration food, water, each other, and the local baboons. A troop's daily decision about where to forage fit inside the larger process that determined where they'd base themselves and how far they would range.

What I did not anticipate was the elephants. You will see how an extraordinary, rare event can blow up the normal processes that determine where a group situates itself, what it does there, and even what it will do in the future.

**September 1994**: I spend a rare night with the Malaika baboons.* They're sleeping atop White Rocks while I'm below, seeking refuge from the cold in my parked car. I can't follow them to the top—that would be a technical climb for me—I can only marvel at their effortless ascent in the evenings. I have chosen a moonlit night so I can see as much as possible. I'm determined to be there when the baboons wake up, descend from the rocks, and begin their daily caucus. As the sun finally peeks above the horizon, most of Malaika is down from the rocks. A few animals nestle together on a ledge formed by two enormous rocks, one balancing on the other. Early morning sunrays reach them there, but also chilly gusts. The rest of the group shelters in a crevice between three rocks, in the shadows but protected from wind. I stand below and point myself eastward into the sun. In the distance, Mount Kenya's outline, snow shining on top, emerges from the morning mist. I take a moment to savor the shades of pink, orange, and yellow. The chill begins to wear off and I'm down two layers of clothes by 8:00 a.m.

Already, individual baboons are trying to convince others to follow them. The options at White Rocks are many since food can be found in any direction. I note who is pointing in what direction and

*Formerly known as the Bad Boys (Wabaya), crop raiders at Kekopey.

the signals they use to mobilize others. I hear grunts and lipsmacks and see "come hither" looks. I check for responses. Today's decision process took about 45 minutes, not as long as some I'd seen. By the end, I had dashed around the base of White Rocks at least ten times, tracking the baboons' negotiations and hoping to parse the moment of decision.

Studying these dynamics was not going to be easy. At first, the best I could do was a running commentary. I felt like an announcer calling a sports event:

> Now Zilla turns and grunts, looking at large male Sharman, but Sharman doesn't see her or is ignoring her signal. Herakles, another large male, tries a different direction but is intercepted by a consort group running past. The speed and number of consort follower males scatters the baboons near Herakles so his suggested direction loses momentum. The consort group runs east oblivious to the rest of the baboons—clearly not a strategic foraging decision, but then these baboons are not interested in food just now . . .

My usual data collection methods fall short. My solution is to treat each troop movement as an organic whole. A specially designed "data sheet" records my best guesses about the interactions that cause shifts in the troop's direction. For example, when a particular animal attempts to lead, I note which baboons are involved in the ensuing negotiations, their types of interactions, and the overall impact on the troop. Once the troop "decides" where to go, I record the foods they target as well as the ecological context.

Were there leaders? That depends on how you define leader. Some baboons were likelier to lead than others, and those possessing *influence* usually prevailed. While hierarchy played a part, it was secondary. I was enthralled by the interplay between individual animals who ventured out front and the rest of the group as they deliberated whether to follow. It was like election day every day. While it would be a reach to claim would-be leaders had humanlike ambition, it looked very much like the young males who raced ahead were saying, “Look at me, I’m awesome.” Perhaps they hoped acting out influence would confer it? Real influence flows through social networks of kinship and friendship, facilitated, and sometimes interrupted, by ever-shifting baboon dynamics. Influence also requires trust, and trust must be earned. The baboon with influence has proven knowledge and competence as a forager but is also socially accomplished. In other words, the influencer is both socially and ecologically astute.

There were always negotiations. Baboons used their communication system—postures, gestures, and sounds—to “discuss” where and when to go. Approaches and avoids, aggression, consort tactics, and agonistic buffering all came into play, often disruptively. I identified many distinct factors that influence troop movements (see appendix 3). Then, suddenly, the conversation ends. Consensus emerges and the troop moves north, east, west or south, even if that wasn’t everyone’s choice. In the end, the group’s decision is what matters most.[1]

I was seeing a fascinating push/pull, a sort of tension, between what some individual baboons wanted to do and what the group decides to do. A similar tension appeared in the ways baboons mediated between the social and the ecological. I wondered why evolutionary arguments didn’t recognize the many ways that living in a cohesive social group limits what a single baboon can do. An animal can be aware of better food elsewhere but bends to the will of the group. This means compromise lies at the heart of their decision-making—except in rare exceptions, like when the young males split from Pumphouse to get farmers’ crops.

But many questions remain. We still don’t fully understand what has clicked into place when a troop suddenly and purposefully moves in a single direction. At the time I delved into the new science of complexity theory offering potential clues. It describes a process of “emergence,” the moment when the whole becomes more than the sum of the parts.[2] Recent studies of bees, birds, and humans have identified "swarm mentality," a similar emergent phenomenon.[3] When a group of baboons coalesces, it's like concert musicians ending their warm-up and hitting the first notes of a symphony. The whole exceeds the sum of the parts by providing the most efficient way to find food in the safest possible way.

Is there a "tipping point?" In 2012, scientist Margaret Crofoot and her colleagues put radio collars on most members of a small baboon troop elsewhere on Kenya's Laikipia Plateau.[4] They argue, based on data collected over two weeks, that troop movement decisions are a shared democratic process. Still, even such a wonderful advance in documenting the complexity of troop movements doesn't tell us precisely *how* that final decision is made. We can identify distinct patterns, but they are diverse and retain an element of mystery.

---

A baboon troop's movements hinged on more than daily decisions. Troop dynamics can be powerfully changed by circumstance. Some years after starting my study of troop movements, I returned from my teaching term and went to check on Malaika. I found them clustered together, like many mornings, but something jarred. There were so few. I had found it hard to believe reports that this troop, already small, had lost yet more animals. Maybe others were hiding nearby? I scoured the surrounding area and found none. The reports were true, and all the sadder because many had been killed by humans, local men who had been incited to kill baboons in the misguided belief that I would hire them. (To say they were mistaken would be an understatement.) Soon after the incident, several males left Malaika to join other troops. One large male remained, adolescent Shasta, in a troop with only 19 animals.

Shasta wasn't yet full size, but his sharp canines had descended, as they do in the muzzles of all males as they grow. Male baboon puberty unfolded between six and ten years, and included those canines, a growth spurt, descended testes, a bloom of muscle mass, and thickened mantle hair. Shasta had blondish hair, a pointy face with a protruding nose, like Rad of earlier days, and gently sloping eyebrows that said, "kind baboon." I didn't think he was ready to assume adult responsibilities, but I would see.

On this morning, the Malaika animals huddled together, grooming, perhaps taking comfort from bodily contact. Last night, they slept alongside a local troop whose home range overlapped theirs. We named that group Soit, "rock" in the Maa language.* I arrived before daybreak to catch early interactions between the groups. Today, Malaika only wanted to get away from the large, dominant troop and dashed into the bush without deliberation.

I returned the next day. Zilla, at 23, is the group's elder and matriarch. (At Kekopey, she'd have been middle aged.) As Malaika left the rocks, Zilla grunted, stopped, looked back, and beckoned her family with several "come hithers." The troop caught up as she headed toward an area we called Central Corm Site. Shasta hung back. Shasta monitored Zilla's movements and she his. Both were keenly attuned to Soit nearby. I saw a Malaika female with a sexual swelling sitting with Soit. Will Shasta stay behind to monitor her? Not today. He rushed to catch up with Zilla.

At the corming site, Shasta climbed a rock to scan for the other troop. I heard his hum-roar-grunt, a unique male call, yet to be interpreted. Zilla was moving toward an area with more tempting foods. Today, the division of labor between Zilla and Shasta was obvious. Zilla led the group to the best food, while Shasta focused on group integrity and protection. He also monitored Maasai livestock and human herders, off to the east.

*Soit would go on to have a major role in Malaika's future, something we had no clue about at the time.

Before the killings and the male exodus, Zilla and Shasta were often indicators* in daily troop movements.[5] Now that their influence was uncontestable. I noticed something else. Never in all my years of watching had I seen anything resembling the "baboon model," yet here it was. Shasta had become the archetypal male "leader." When the group sensed danger, females and their young clustered near him. Yet Zilla's influence was unmistakable and ran counter to the male-centric baboon model. Were these anomalies a function of the group's losses? My only certainty was that events beyond their control had altered the group's behavior and streamlined their movements. I was already thinking about the evolutionary implications of aberrant circumstances. Here was evidence that a group's way of conducting itself could be altered by chance occurrences.

---

The daily wanderings of a baboon troop occur within a particular range, an informally bounded area that a troop mostly keeps to but does not defend. It can be a few square miles in a food-rich forest or more than 10 square miles in arid terrain. Adjacent baboon ranges are a bit like Venn diagrams, circles with overlapping borders. Each troop has a home range with "core areas" that include preferred sleeping sites and nearby spots for feeding and hanging out. Security-seeking females enforce the boundaries, feeling safer in known areas and anxious outside them. Males, meanwhile, circulate through several troops in their lifetimes, gaining useful knowledge about different areas and home ranges. But Pumphouse and Malaika were mapping new ranges from scratch when they suddenly found themselves translocated. Both soon began cautious

*Indicator is a term adopted by Hans Kummer, who earlier studied hamadryas baboons, to describe baboons who routinely attempt to lead troop movements.

exploration of their new home. As they became familiar with foods, sleeping sites, and the home ranges of other troops, they carved out their own separate ranges.

Once settled, Malaika seemed content to stay near their release site, drifting only slightly to the west, east, or north. Pumphouse was very different. By 2000, 16 years after the translocation, this troop had gradually ventured eastward to the edge of the Mukogodo Forest, more than 13 miles (as the crow flies) from their release site. I suspected that this radical range shift might have been due in part to the fact that they lacked historic roots in this new place. But there were other factors as well . . .

Though females tend to resist forays into unknown territory, Pumphouse found itself in a new range around Bridal Rocks, after an episode I'll soon describe. A local troop we called Bridal Troop often slept there as well. The two troops kept their distance but monitored each other, even though youngsters sometimes mingled and adolescent males eyed neighboring females. Each troop slept in its own space on this large kopje.

One early morning at Bridal Rocks, I noticed that Thistle, Peggy's youngest granddaughter, had disappeared. She'd had a good-sized sexual swelling the last time I saw her. I was about to launch a search when she reappeared with a Bridal Troop male in tow. For the next few days, the couple moved back and forth between the two troops. Thistle stopped visiting Bridal Troop when her swelling deflated, meaning she'd ovulated and was no longer attractive or interested. To my surprise, her male partner stayed with Pumphouse, becoming the first of eight Bridal males to transfer to Pumphouse. This made sense since males normally transfer and Pumphouse was conveniently close. Then—perhaps because of these new ties between the two groups—Pumphouse began to follow Bridal Troop on daily foraging missions. Bridal, after all, had been there longer, ranged more widely, and knew where the food was. But when the local group ventured beyond the area familiar to Pumphouse females, Pumphouse put on the female brakes, as if an invisible barrier had gone up.

What Pumphouse was missing out on was Windmill Gully, where a seasonal river supported a bonanza of baboon foods. Soil washes into the gully during the rains, leaving a deep, rich bed of earth favored by fever tree acacias. As important, deeper soils hold moisture longer, allowing rainy season foods to last into the dry season. Here you could find green grass, grewia and cordia trees with berries, lyceum bushes with soft leaves and flowers, and more, long after they'd disappeared from the plains and slopes above.

It took many months and the concerted efforts of Bridal immigrant males to entice Pumphouse to Windmill Gully. A single male couldn't do it, because as a newcomer he lacked "pull" with Pumphouse females. Only the combined social networks of several new males finally overcame the troop's strong natural resistance. Once Pumphouse females saw the bounteous food at Windmill Gully, they were all in. Did those males have a plan? I doubt it. They knew where the best food was, but they'd bent to the will of the more cautious Pumphouse members. That changed as more immigrants joined and made friends with Pumphouse females. Because those males persisted, all benefited and Pumphouse expanded its range.

Pumphouse continued to move east beyond Windmill Gully. I realized that this incremental eastward movement had two precursors. First, a critical mass of well-traveled immigrant males needed to join Pumphouse. Second, those immigrant males had to establish social networks and gain influence. Only then could they goad Pumphouse into exploring new terrain. The move from Windmill Gully to Fig Tree Rocks, another baboon sleeping site to the east, happened this way. Finally, even Windmill Gully was left behind.

---

How Pumphouse wound up at Bridal Rocks is another story. Two years after the translocation, a decisive event occurred that changed not only the Pumphouse range, but my thinking about evolutionary calculations. This was the night of the elephants.

**August 1986**: Dawn is breaking. I climb to meet the Pumphouse baboons before they've made their way down from Geoffrey's Lookout, where they slept. I expect it will be a routine day. Once at the base of the kopje, the baboons will have a brief round of socializing before heading off to forage. Geoffrey's Lookout is a favorite because of its rocky pinnacle that safely towers over the plains.

On my way home the night before, I'd passed a large herd of elephants in the grassland below. I took note. Here on the savanna, fear mixes with awe when it comes to elephants. The popular image of elephants trumpeting and crashing through the jungle is misleading. Elephants can be almost catlike as they wend across the savanna. If you're near, you may hear branches break or a soft rumble, as I did one night when I woke to see an elephant rubbing itself in the moonlight against the outer wall of my bedroom. In any circumstance, elephants require vigilance; they can turn aggressive on a dime if, on their daily rounds, they are surprised by a perceived threat.

At Geoffrey's Lookout, I see no baboons. I check the far side of the kopje. No baboons. I search the surrounding area, thinking maybe they'd risen earlier than usual. Still no baboons. How can this be? But in my hunt for baboons, I find a clue: footprints of elephants. No doubt the herd I saw yesterday. From the looks of their tracks, the elephants came up the trail to the sleeping rocks, milled around, and left. I imagine the baboons were safely tucked away on top of the kopje when the herd paid its visit. I pick up a trail of fresh baboon footprints. They lead me away from the rocks and past the only sign of humans in this area, a small police post. A few uniformed officers are stationed there, remnants of a time when this peaceful part of Kenya was threatened by bandits coming from the "wild North." The officers confirm that they heard baboons pass by in the night and thought it "strange." An officer points at distant Bridal Rocks, situated at the outer edge of the home range charted by Pumphouse over the past two years. I raise my binoculars for a closer look. There were the baboons, resting and sunning themselves.

Especially at times like this, I wish baboons could talk. Still, I could easily imagine the baboon perspective. Baboons, like most primates, find nighttime dangerous, with good reason. Leopards silently prowl at night and, unseen, strike with lightning speed. As for elephants, they were a novel and unsettling feature of life here. Local baboons mixed comfortably with them, but the translocated baboons never saw them on Kekopey and had yet to relax around them. *I* would feel a surge of panic if an elephant herd massed around my home. Baboons probably would as well.

This chance occurrence resulted in a quantum range shift eastward. Pumphouse abandoned more than half of its home range, the part used by elephants. The impact was far-reaching. Males and even females with babies—who seldom leave the troop of their birth—left to join local troops.* Pumphouse declined from 60 to 22 individuals, and I worried it might be absorbed into the local baboon population. Thankfully, new males began to filter in and revitalized the group.

Unlike daily activities, rare events can't be quantified, but their reverberations can be stunning. The baboon-elephant encounter was an "accident," a happenstance, yet for years the group avoided the elephant area. This was the ecology of fear, which often displaces animals from the best foraging grounds.[6] What I realized was that any group's history would contain events that altered its trajectory and helped mold its future. After all, today doesn't start from scratch; it carries echoes of history. The elephant encounter joined my list of rare events along with the translocation and

*We counted five distinct local groups, several of which we also studied because of their critical contact with translocated animals.

Cripple Troop's reunion. These kinds of singularities add another layer of complexity to baboons' lives and to the task of understanding them. Clearly, it seemed to me, these events should not be ignored but factored into the understanding of a group's behavior.

The baboons had given me a scientific bonanza. My biggest surprise was how much history matters. Recall that Malaika and Pumphouse were once one group and shared a genetic heritage. They faced the same challenges after translocation. But the troops' fates diverged due to the elephant encounter. Chance and happenstance, not just adaptation, create history. History fashions contingencies that facilitate or constrain the future actions of individuals and groups.

Clear distinctions emerged when we compared different Pumphouse range shifts. With the elephants, it was a "push." Then the troop hopscotched east, a "pull," spurred by the indigenous males from Bridal Troop. The result was better food and new knowledge. Watching that arduous process, I learned how hard it is for baboons to share information without the benefit of language.

I can only speculate as to the relative importance of ecological and social complexity, and I do in chapter 15. But I feel certain about a particular insight gleaned from this study of troop movements and home range shifts. Today, baboons (and humans) must simultaneously navigate the social and the ecological in a world where the only constant is change.[7]

Thinking about how to interpret baboon troop movements, I revisited what I'd learned about baboon ways. Daily life consists of many intricately entangled elements, both social and ecological. I had already recognized social complexity; now it became socio-ecological complexity. Baboons make existential choices in an ecological context. The raiders did, and so do other baboons who break away to form new troops. Many decisions involve a complex interplay between the social and the ecological.

Yet a current key evolutionary assumption is that reproduction is the sine qua non of individual success. I could not reconcile this view with my observations. What I saw was the social group supporting survival, *not* an individual's strength or dominance, not even in times of scarcity. Competition for limited resources, rank, or reproductive success mattered less to baboon success than collaboration and shared knowledge. Evolutionary principles certainly apply to baboons, but measuring only reproductive success captures just a corner of the baboon tapestry. You would miss a myriad of other features of baboon life that undergird the accomplishments of individual animals and the group as a whole. That includes the way baboons manage the deeply interconnected social and ecological realms. Navigating both simultaneously may be the greatest challenge of baboon daily life, though it is what they do—naturally. Individual baboons are much more than mere "gene survival" machines.

Evolution wired animals to seek the best available food. But this evolutionary rule, and others, operate in a context of unimaginable complexity, and they're modified by contingencies such as history and chance. In our world, such contingencies often overwhelm evolutionary determinism.[8]

Once again, my thoughts turned to outcome and process. Evolutionary predictions focus on the outcomes of behavior but neglect the processes that achieved those outcomes. The baboons forced me to think about *process*. Even when baboons act as evolution would expect, ignoring process or dismissing it as irrelevant can lead to wrong interpretations. That elephants visited Pumphouse, not Malaika, was a chance event resulting in divergent sets of constraints and opportunities for each group. My fancy home range grid maps showed Pumphouse moving east in stepwise fashion, while Malaika barely budged. The grids gave me scientific credibility, but they didn't explain *why* the two troops' movements were so wildly unalike.

What to do with these impressions? I didn't want to create another "just-so story"—a speculative, though possibly attention-grabbing, evolutionary explanation of the two outcomes. I stepped back and took a broader view. What the baboons had shown me and convinced me of was that few things are inevitable. This put me at odds with the current evolutionary predictions of genetic inevitabilities and predetermined behaviors and outcomes.

The baboons also remind me that every fresh set of circumstances—fortunate or not—creates opportunities to discover new things about them.

---

My family grew while Pumphouse shifted their home range. Carissa, my daughter, was born soon after the translocation. I took her to see the baboons when she was a toddler. There was no mistaking the mutual fascination. They looked at her; she looked at them, then she raced toward the nearest baboons. I ran interference before we could find out what interspecies play would look like. Guy was born two years later. I thought Carissa was a good baby until Guy was born. He was exceptional, but unlike with Carissa I didn't treat him as a subject of my notes. Jonah and I traded off taking care of the children; I stayed with them when he went to Amboseli, and he stayed with them when I went to the baboons. Sometimes he brought Carissa and Guy to Chololo at the end of my field stint. I was disappointed that as they grew neither showed much interest in the baboons, preferring to climb kopjes or look for spent bullets left behind after Kenya police and British Army exercises. I never expected either to follow in my footsteps, but I hoped one day they would appreciate how special it was to see a wild group of baboons up close. Eventually they did, and both made interesting lives for themselves in Africa.

I found myself on a path I never intended to take. And, like the baboons, I ended up here because of chance encounters and accidents in my history that forced me to look in new directions. Unlike the baboons, who could rely on social support when they found themselves in a strange new land, I was on my own, making statements about baboons that few colleagues took seriously. But it didn't make me any less certain that the baboons I watched led lives that contained options, choices, decisions, unexpected events and outcomes, and the residue of historic experiences. This was a far cry from a genetically ordained, predetermined existence proposed by basic evolutionary metrics. The baboons had turned my intellectual world upside down because I'd seen much in baboon daily life that was not predetermined.

But I had other worries. Now that Malaika was so diminished, its odds for survival were looking grim.

Chapter 9

# Mergers and Acquisitions (1999–2001)

**SEPTEMBER 1999:** Another early morning at White Rocks. My annual teaching quarter behind me, I'm glad to be back in the field. But what greets me here is concerning. Malaika has lost yet more animals in my absence; only 13 baboons remain. I'm told predation has taken its toll. Earlier, the predators had been humans. Now they are the baboons' natural foe, the leopard. Today they share sleeping rocks with Soit, a much larger group with about 70 members. Soit seldom sleeps at White Rocks, having many sleeping sites to choose from in their vast home range. Shasta, who had so vigilantly guarded Malaika after its initial decimation, is gone and is missed, at least by me. I note that several Soit subadult males must have joined Malaika when they were in adjacent troops, but it's still very small as baboon groups go.

Here at the base of the rocks, Soit and Malaika look like a single troop, but odd behaviors tell me they are actually two. The males in both groups are visibly tense—jumpy and intent on keeping "their" females from the neighboring group. This is not an easy task. Malaika's three adult females are keen to socialize with Soit's many little black infants, with no babies to fondle in their own small troop. One male in Soit, Nex, stands out in the commotion. He vigilantly patrols Soit's boundary, successfully blocking both Malaika and Soit females who try to cross over.

Since I know these animals by sight, I can see that Malaika's social networks remain intact and separate from social subgroups in Soit. It's reminiscent of middle school cliques at recess: Small groups cluster close to each other but never mingle. There's more evidence of separation when it's time to leave the rocks and head out to forage. Confusion reigns. Soit has its own indicator animals—the ones who run out front and attempt to lead—as does Malaika. But Malaika wants to stay close to Soit, which causes a strange-looking stalemate. The two troops finally leave, at roughly the same time, but separately, with Soit indifferent to Malaika's actions.

These troops have different but overlapping home ranges. Malaika's is small compared to Soit's. This presents Malaika with a dilemma. On one hand, it seeks to avoid direct contact with Soit, the dominant troop. On the other, it seems to value some degree of closeness, possibly as a security measure. But if Malaika wants to maintain proximity, they'll need to leave their home range and follow Soit into unknown territory. I'm very interested to see what they do.

Still within Malaika's range, the two troops stumble their way through a foraging foray. Then Malaika suddenly sits down and refuses to move. Soit keeps going, heedless of the growing separation. Still, Malaika does not budge. This is the border of Malaika's home range. As I learned during the Pumphouse home range shifts, females resist leaving familiar terrain, an expression of their natural caution. Malaika waits, along with me, feeding on what's there and socializing. Soit returns, hours later. Then Malaika falls behind the bigger group, shadowing them on the way back to White Rocks. At the end of the day, Soit ascends to the highest rocks and settles in for the night. Malaika hangs back, waiting until it's almost dark to climb to the lowest of the three rocks—displaced by Soit from their preferred pinnacle rocks.

What I'm seeing here is highly unusual: social tension between two unbalanced troops. I mean to track what's happening between them, but I have a problem. I had been diagnosed with scoliosis as a child but managed well enough until 1992 when my back began to fail. I was paying the price of mature motherhood. Each of the two births softened my pelvis and destabilized my spinal curvature. By 1998, I'd lost four inches in height and my spine looked like a backward S. Standing, walking, and even sitting are now difficult. To no avail, I will my muscles to support me without the benefit of an effective bony structure. The pain is relentless. I'm obliged to take strong medications just to function. The bumpy ride out to Malaika taxes me, and my days with the baboons end earlier. But giving up is not an option. I strip my activities down to the essentials: I watch baboons, I teach, I parent. I carry cushions to sit on, wear a thick back support belt when I walk, and invent a pillow top for my mattress so I can sleep.* I arrive at camp carting my new equipment.

---

Social living is a key evolutionary adaptation in the Primate order. The presumed benefits are several: protection from predators, mate availability, socialization of the young, and a competitive advantage in the search for food. The group has shared knowledge of resources and potential perils, more than any individual could possess. Yet no one asks the most basic questions about a primate group. What forces actually bring about a group's formation and,

*This I wish I had patented. Similar products would later hit the market.

once formed, how is the group's integrity maintained? As scientists, we've taken the group for granted as if, prima facie, it's a timeless structure. Primate babies are born into a group and perpetuation results, meaning studies begin and end with a group already in place. But stop and think about it. That group is, in fact, an *outcome* of a *process*.

I first mulled these questions in the 1980s, during my collaboration with Bruno Latour, the noted French intellectual. Bruno's investigations spanned sociology, anthropology, and the philosophy of science and happened to mesh with my interests.[1] (In chapter 14 I describe our work together.) Now, years later, I was seeing the abstract made concrete as our ideas about social groups played out in real time. Malaika and Soit appeared to be fusing, and the process was anything but simple. Over the next two years, I would track this merger with all its shifting dynamics and surprising turns. It was something I'd never seen in 30 years of baboon watching, nor had anyone else. By its end, what had seemed inexplicable—the reasons behind the fusion and its painfully halting progress—made sense, historically, ecologically, evolutionarily, and behaviorally, in a uniquely baboon way.

I'd seen Pumphouse split during the crop-raiding period. Fissions—when a baboon troop breaks apart—are infrequent but expected under certain circumstances, as when a troop grows too large.* Baboon fusions, or troop mergers, had been seen only twice, once at Amboseli National Park in the 1960s and once at Mikumi National Park in Tanzania in the 1990s.[2] Fusion is rare even among other Old World monkeys. Only four appear in the literature, two among vervet monkeys (*Chlorocebus pygerythrus*),[†] one among toque

*However, few troop splits are described in the literature.

[†]The vervet monkey fusions were observed in Amboseli National Park, Kenya, and another in Laikipia, Kenya. The toque macaque fusion occurred in Sri Lanka.

macaques (*Macaca sinica*), and one among Japanese macaques (*Macaca fuscata*).[3] In these recorded cases, fusion happens quickly, usually after two groups meet at a sleeping site in the evening and leave together as one the next morning.

A distinction needs to be made with another category of primate group: the primate fission/fusion society. In several species, including chimpanzees and spider monkeys,* normal social life involves fission and fusion in the form of repeated splits and reunions over the course of a day.[4] Baboon group structure is stable by comparison, making what I was seeing highly unusual.

**September 2000**: Dawn at White Rocks, a year later. It's a cold and windy morning, so I bundle up. My thick back brace gives me extra warmth. These days I can't sit, not on the ground and not on the rocks. The baboons come down from the top of the kopje where they slept and find shelter from the wind in the shaded lee of the rocks. A few animals move onto the sunny ledge, facing east. I follow them, happy to feel the sun's warmth. But I need to keep an eye on all, so I'm back and forth until both troops have descended to the base of the kopje.

The fusion has modestly progressed in the last year. Troop movements remain chaotic because both Soit and Malaika's indicator animals still run out front and cause indecision. Things aren't much better after they leave the rocks. All move in the same direction but maintain their group boundaries. Up to this point, Malaika had appeared to be driving the fusion dynamics. They track Soit, staying as close as possible but deal as best they can with Soit's tendency to venture beyond the smaller troop's home range border. Today, I am baffled when Soit won't allow Malaika to drift too far away. This is new.

I hadn't seen much competition for food, neither during the translocation nor in the years that followed. So far, there'd been little

*Genus *Ateles*, a New World monkey from South America.

direct feeding conflict between these two troops. But as acacias entered their bloom cycle, Soit doubled down on its dominance by pushing Malaika from the few flowering trees. Acacia flowers are an important staple in the dry season. A baboon can eat 300 of the tiny, cream-colored blossoms in one sitting and then, after a respite, eat still more.* The appeal is probably the high sugar content. I'd sampled a few at Kekopey and experienced not only the sweetness but the strange desiccating effect of the flowers in my mouth. No wonder baboons always looked for water after an acacia flower feast.

On this occasion, Malaika, kept from the acacias, was able to slip away to forage on its own. Soit initially failed to notice Malaika's absence. When they did, the Soit baboons set out in search of the errant troop, found them, and mobbed them. It looked like Malaika had been punished for straying too far.

I'm still trying to make sense of what's going on between these two troops. Now Soit, not just Malaika, seeks closeness. I note that the last year has seen a change not only in Soit's behavior but in its numbers. Predation has trimmed Soit from more than 70 animals to only 35. This may be significant.

---

**February 2001**: Dawn at White Rocks, six months later. Dawn is my favorite time of day, despite the chill, because I can get as close to the baboons as they are to each other, which is very close.† Only a few feet separate us. With my limited mobility, I stay around clusters of baboons below the kopjes and avoid scaling the rocks. I begin my early morning notes. Something is different. The usual hard

*The flowers, about the size of the tip of my little finger, look like fluffy puff balls.

†This was before the 2021 IUCN Primate Specialist Group restrictions on photos or videos taken less than the specified distance required between humans and nonhuman primates. They hoped to discourage images that fed into the primate pet trade.

boundary between the two troops is gone. Females from both sides enjoy visiting each other. A Soit male, Caterpillar, is surrounded by a group of Malaika females. He'd joined Malaika as a juvenile then returned to Soit as a middle-aged adult. His old Malaika friends seem glad to see him.

As the sun rises, mingling continues. I soon discover why everyone seems so relaxed. Nex, the aggressive micromanager of Soit females, is gone. Without his vigilant enforcement, the physical border between Malaika and Soit has blurred. Could so much depend on just one baboon? I wondered if he'd worn himself out and moved on. With Nex out of the picture, the fusion has rapidly progressed. Relationships have developed across group lines, between females, and between males and females.

Two years ago, I had a hard time imagining that Malaika and Soit could ever be a single troop. But here they are, sleeping, resting, and socializing together, no longer self-segregated. I make notes scanning across the baboons. Malaika's Rebecca, Naomi's great-granddaughter, is as likely to sit next to Soit's Topic or Scotty as she is to her Malaika troopmate Heather, who happens to be Zilla's granddaughter. When the time comes to leave the sleeping site, chaos has given way to a more orderly set of negotiations. The terrible La Niña drought of 2000 has abated, but food is still in short supply. Both troops have retained their own indicator animals, but it's a normal sort of negotiation as they point in different directions, not frenetic, the way it was. Finally, Soit wins out. As the day wears on, I see a harmoniously merged group with members that groom each other, forage together, and visit each other's babies.

Since I'm doing science here, I need criteria that will allow me to quantify stages in the fusion process and determine its completion. But I have none, because the few previously recorded mergers were overnight affairs, not spread over two years. Once again, I need to be creative. Both fission and fusion result in groups, so I figure I can apply my own fission benchmarks to chart this fusion. My repurposed checklist has 15 criteria (see appendix 4). Among them are measures of group integration during activities such as troop movements and foraging, who's mating with whom, who's friends with whom, and who gets access to whose babies. By August 2001, two years after the two groups began their fraught association, Malaika and Soit meet all the criteria of a single troop. I declare them fused and give them a new name: Nabo, which means coming together in the Maasai language. That is just what they did.

---

There is a backstory to this troop merger that I figured out retrospectively. This is how science in the wild works. You document an unusual event, and later you try to make sense of it. I have spoken of how history matters, along with ecology, social behavior, and evolution. Drawing on my earlier findings and thoughts about these aspects of baboon life, I approached the unification of Soit and Malaika. Of the many questions I had about the fusion, two loomed largest: Why fuse, and why now? I turned to the evidence I'd gathered: quantitative data sheets, natural history reports (see chapter 16), and notes taken while speaking with local residents. And I began to consider which evolutionary principles might apply.

Two events stood out as critical. The first was ecological: extreme weather. In the years preceding the merger, 1997 and 1998, El Niño rains pounded the savanna.[5] El Niño had different impacts around the globe. In Kenya, rain fell harder and longer than at any time in memory. Our project recorded four times the normal rainfall in 18 months. The tracks and roads to the baboons turned into nearly impassable bogs, and the bogs hid patches of dreaded black cotton soil. If you chanced to step into one of these patches, your shoes would be freighted with a pound at least of the mucky stuff, after just a few steps. To free a car mired in black cotton soil you would need a shovel, rocks, and several people willing to push. If you were alone in that car, you'd be walking the rest of the way. Once you got to the baboons, in the rain, you sought whatever tree or rock offered a modicum of protection, then observed huddled baboons as your data sheets went limp. In normal rainy seasons, showers are intermittent and baboon watching is usually possible. El Niño rains were different. I spent a lot of time at camp working on data.

But there was an upside. The El Niño rains produced an explosion in plant life. Plants appeared that the baboons and I had never seen. The baboons added many to their diet.

The second event was the creation of a wildlife sanctuary in 1996, a historic occurrence. The sanctuary overlapped with both Malaika and Soit home ranges and included White Rocks. Wildlife, sensing a new zone of safety, spilled over from neighboring ranches and into the sanctuary. Here they found food as well as water in a newly constructed dam. I was soon seeing greater numbers of impalas, Grant's gazelles, zebras, giraffes, gerenuk, elephants, and dik-diks on my way to the baboons.

Inevitably, these prey animals drew predators: lions, hyenas, the occasional cheetah, and, most relevant for baboons, leopards. One rarely catches a glimpse of a leopard, the lethal night-roaming phantoms of the savanna. But we speculated that when a healthy baboon mysteriously vanished overnight, it was likely taken by a leopard. Malaika and Soit were now surrounded by wildlife, including predators that fed on baboons.

Baboons were seemingly exempt from predation during this time, but not during the devastating La Niña drought that soon followed. Most of the prey species left the sanctuary in search of food, but the predators stayed. We began to hear reports of leopards taking goats from Maasai corrals (*bomas*). When the bomas were secured, the leopards switched to eating the pastoralists' dogs. When all the dogs had been eaten, the leopards started on the baboons. Baboon numbers were declining by the year 2000, as we knew from Malaika and later from Soit.

---

Using evolutionary theory, I considered the fusion from each troop's point of view. The group had evolved to provide security, with many animals poised to sense a predator's approach. A small group like Malaika would be less well equipped to provide that security. It seemed no accident that Malaika started following Soit after leopards had reduced their numbers to only 13 individuals. When Soit ranged away from them, I could see a shift in Malaika's behavior from relaxed to edgy and anxious. They clearly felt safer when Soit was nearby.

Soit, on the other hand, had little to gain from Malaika, at least initially, because of its robust numbers. Food abundance brought by the El Niño rains resulted in a baby boom of 20 infants. Then came the La Niña drought and extreme food scarcity. Soit began to lose animals. The babies all died. Then the baboons became leopard prey. When predation reduced Soit's group to only 35 animals, I observed Soit's behavior toward Malaika change. Soit began to focus on Malaika, even demanding closeness. My hunch was that Soit sensed that a larger group is a safer group. Thus, Malaika's proximity gave Soit added protection, and also, perhaps, animals to be picked off by leopards in lieu of their own.

The benefits for Malaika were clear, but there were costs as well. By virtue of size, Soit remained the dominant subgroup throughout the merger. That gave its members the edge in feeding competition, to the disadvantage of Malaika, especially after Soit began to insist on Malaika's proximity. Fortunately, as I had seen, Malaika was on occasion able to slip away from Soit to feed in peace and avoid constant displacement. Yet at the first hint of danger, Malaika rushed back to Soit for safety, in a trade-off Malaika was willing to make.

Then I considered the female dominance hierarchy. At first, after the fusion, it looked like the top Malaika females would keep their ranks. They certainly tried. High-ranking Zilla and her cousin Desiree refused to let Soit females displace them and pushed their way up the Soit female hierarchy, backed by their Malaika male friends. Zilla and Desiree seemed to relish playing "mean girls." When a lone Soit female found herself in a Malaika feeding group, Zilla and Desiree were quick to harass and displace her from her meal—but only when other Soit animals weren't around to defend her. But as their Malaika male friends drifted toward Soit females, Malaika females lost needed support and fell in rank. The lowest-ranking Soit females were especially resistant to Malaika females' aspirations, vigorously harassing and displacing Malaika females at every opportunity. These Soit females had the most to lose. If they conceded rank to Malaika females, they'd find themselves on the lowest rungs of a large troop—the first to be displaced from feeding spots and in everyday social interactions. Eventually, the two female hierarchies knitted together. All Soit females ranked above all Malaika females in a two-tier system, with females retaining their relative troop rankings in the new order.

Soit females had a different set of costs to benefits. They kept their feeding advantage since they could displace any lower-ranking Malaika female from food. But there may have been less food to go around due to an unexpected outcome of the fusion. Recall that earlier, Malaika females refused to venture outside the border of their home range. After fusion, as Nabo, I mapped their new range. It was a compromise—larger than Malaika's old range but a good deal smaller than Soit's formerly expansive range. Malaika's net benefits were more food and new sleeping sites. Soit had less food and fewer sleeping sites but retained their dominance.

Males had their own story. During the merger, Soit males consorted with both Malaika and Soit females. Wiggle was an interesting exception. The top-ranking male in Malaika, he had landed in the upper echelon of the Nabo male hierarchy. Yet, for a while he paid little heed to Soit females, preferring the company of his former Malaika female friends. Indeed, when Wiggle eventually turned his attention to Soit females, I knew the blending was complete. But as the males sorted out their places in a new hierarchy, I realized that I was watching males behave almost as they normally do in a stable troop. They test each other, they bluff, they compete for females, and they seem to accept, often grudgingly, that one day you're "in," and the next you may be "out." In a sense, it was a more organic process than that of the females, whose struggles for a higher rank in a fixed hierarchy had lasting consequences. Indeed, the process of male hierarchical integration in Nabo provided yet more evidence of the relative dynamism of male rank, and yet another strike against the "baboon model" (see appendix 4).

Over the next several years, Nabo grew from 54 to 69 individuals. compared to Soit's previous 35 animals and Malaika's 13. As evolutionary theory predicted, they experienced less predation under all ecological conditions. One might think that leopards had left the area, but sightings continued per reports from the community. The larger group did seem to offer greater protection, something seemingly sensed by both Soit and Malaika as they negotiated their merger.

Serendipity handed me a de facto control group in the form of a small local troop that foraged near Nabo, the fused troop. Though I didn't know their history, I saw them often enough to document their fate. Unlike sizable Nabo, this troop experienced a steady loss of males, females, and youngsters during and after the Malaika/Soit fusion. Since the remaining animals looked healthy, I could reasonably surmise that predation accounted for their attrition. This was good *circumstantial* evidence that there is safety in numbers. Bear in mind, this was a test of evolutionary theory, not common sense.

---

I now understood why the fusion took so long. Soit and Malaika were something like East and West Berlin after the fall of the Berlin Wall. Each troop had its own social system and merging them required a massive amount of work, by which I mean negotiation. Two distinct hierarchies needed to mesh, a process that rippled through friendships, consorts, troop movements, and every other aspect of each baboon troop's social domain. I guessed that this fusion was different from the few others on record because Soit and Malaika were both viable troops. Reengineering two viable entities into a single whole is much trickier than forming a patchwork group from remnant animals whose social structure is weak or nonexistent, like the few fusions I'd found in the literature.

---

I hadn't quite appreciated the absolute irreplaceability of the social group in a baboon's life until the translocation. The Nabo fusion deepened this appreciation and added dimensions to my understanding. In the 1960s, when I was first learning about primates, we assumed "being social" and living in a social group was a given. We ticked off those benefits: security, feeding advantages, mate availability, and the passing down of "group" knowledge.

Yet, from the baboon perspective, to say the social group is just about security or feeding feels almost absurdly simplistic. There is so much more going on. Individuals and groups make hourly, daily, and monthly adjustments as they decide where to go and what to do. Males endlessly gauge their standing in the group and debate whether to take action. Females choose among prospective friends and mates. It seems quite clear to me that an action taken by a baboon is not just an outcome of a long-ago evolutionary process. A baboon's behaviors are also the result of decisions it makes as it manages the complexities of everyday life in the context of a group.

I also saw more clearly the constraints of group living. For instance, Zilla may decide to go one direction, but she won't go by herself; she must mobilize others in Malaika to follow. In Nabo, Zilla doesn't have the same influence. Her family is a small part of a larger troop, and she is no longer the top female or even the oldest with the most feeding wisdom. I seldom thought about individual actions as compromises because they occurred so seamlessly.* Yet most actions are an individual's concession to the group's goals.[6]

The fusion showed me that the *entire* social group is negotiated in a process that demands an enormous amount of social work. The raw material of kinship, friendship, rank, sex, knowledge, and strategic behaviors must all be completely reconfigured. My early thoughts about baboon negotiation focused on social strategies that provided alternatives to aggression but needed social relationships to be created and managed. Later, crop-raiding demonstrated that even foraging decisions are negotiated. Troop movement dynamics gave me more data about how negotiation happens. Over time, I was forced to enlarge the context for negotiation from the social to the ecological, and from two-partner exchanges, the basis of most scientific studies, to interactions between multiple individuals.

The toughest negotiation between fused Malaika and Soit clearly occurred between the two female hierarchies. Though I already knew the female hierarchy was a good deal more stable than the males', I had yet to deeply consider its social and evolutionary implications. But when a group of Pumphouse females mounted a stunning insurrection, it was time to plumb the mysteries of the female order.

*Discussions of individuals in a primate group ignore the constraints the group imposes on individuals.

Chapter 10

# The Power of Predictability (1972–2008)

BACK AT KEKOPEY, Peggy and her family were the North Stars of Pumphouse and my early investigations. It was Peggy who first opened my eyes to the hidden female hierarchy. Later, I would uncover its importance. For that I am grateful.

Peggy was the Pumphouse grande dame—the highest-ranking female with the most influence. From her, I learned what it means to be top female, and I tracked her line through generations: her daughter, Thea, followed by her daughter Theodora, then Theodora's daughter Tootsie, and, finally, Tootsie's daughter Kilo.* Each showed me how females of the same rank can be very different despite shared genes, with "personalities" that ran the gamut from feisty to mellow.†

*We ran out of *T* names and switched to *K* names for her matriline.
†The word "personality" was a scientific no-no then; now less so.

Peggy was calm, composed, even generous, seldom claiming the prerogatives of high rank. In primatological circles, primate societies with strong female hierarchies are often described as "despotic," and those without as "egalitarian."[1] I never liked that model or those terms because individuals like Peggy were clearly *not* despotic. In Peggy's world, each baboon knew its rank and rarely had to be reminded. Rank's benefit was most obvious when the troop ate "corms," the underground storage organs of "onion" grass so vital to Pumphouse in the dry season. Peggy's high rank entitled her to displace and take food from a lower-ranking female who'd painstakingly prepped a corm harvesting site. But Peggy sometimes didn't. When she did, her manner was calm, almost genteel. Unlike that of Thea, her adolescent daughter, a flagrant wielder of rank's privileges. She would bully and chase low-ranking females to get their corms when a polite approach would have worked just as well.

Foraging wasn't the only arena where a female could pull rank. Black-colored babies, called black infants, are a great attraction, and Peggy loved to handle them.* But there was a catch. A lower-ranking mother with a black infant would often deferentially back away from Peggy, thus depriving Peggy of a visit with her baby. In those instances, Peggy would quietly follow the mother, making reassuring sounds and gestures. If her blandishments worked, Peggy would lift the baby high in the air, toward her one good eye, then gently set it down. No so Thea. Bullying got her corms, but it backfired when she wanted access to a baby. Her strong-arm tactics often caused a gang of baboons to rush to the baby's defense. Peggy seldom intervened, but she did when Thea unleashed a furor too big to ignore.

Theodora's mild temperament mirrored that of her grandmother, Peggy. Tootsie, meanwhile, shared her grandmother Thea's

*Reminder: Infants are born black with pink faces and only later turn the adult olive color with dark faces.

combative bent. Whether these differences were more deeply rooted in nature (genetics) or nurture (socialization) was something I wondered about but had no way of knowing.

Because Peggy has been so special to me, I keep her skull on my office shelf. That I even have her skull is unusual, due to an incident I wish had not happened. Thoughts of it still make me teary. I had been away for a few days when, in the mid-1970s, I came upon Peggy at the base of the troop's sleeping site. I was distressed to see that her broken leg already had maggots. I wasn't sure she would tolerate my touch, but she stayed calm when I picked her up and carried her to my car. I had never handled a baboon, and she'd certainly never been this close to a human. She remained placid during the two-hour drive to the Institute of Primate Research in Tigoni, and she didn't struggle when I gently lifted her from the car and carried her into the veterinarian's office. But she couldn't be saved. Her injuries—she had probably fallen from the sleeping rocks—were too grave. I knew she would have recognized me, but she may not have sensed that I meant to help. That didn't stop how I felt about her.

---

**Pumphouse, September 2008**: Pumphouse slept at Mission Rocks, more than 13 miles east of their original release site. These rocks—four outcroppings, each suitable for baboon overnights—jut from a ridge that faces south toward Mount Kenya. With my bad back, I'm not exactly an alpinist these days, so I wait for the monkeys to descend and begin their early morning socializing. When they do, a perplexing dynamic unfolds. A coalition of females from the O family starts hounding Buffalo, a young adult female from the current top-ranking family.* The O family is a large, middle-

*Buffalo's family ranked just below top rank at Kekopey but became highest in Pumphouse after translocation, when Peggy's family had few members and fell to second ranking.

ranking matriline descended from Kate and her daughter Olive, subjects in my first study. The scuffle may have started up top; I saw no inciting incident below. What's odd is how intensely the O's are mobbing poor Buffalo. Females rarely behave this way, and usually only in resource disputes between troops (chapter 11) or when an errant male frightens a youngster or female (chapter 2).

The O's egg each other on with eye signals, sounds, and gestures, looking around to see who is with them and who isn't. Buffalo needs the support of her own large, high-ranking family, and I'm not surprised when Dusty, her aunt, shows up. The shocker is that Dusty joins the O's *against* Buffalo instead of defending her niece. The melee has the appearance of an asymmetrical fencing match. The many O's rush Buffalo, Buffalo retreats, then Buffalo pushes forward, only to be repulsed again. This goes on for more than an hour.

You may be wondering how this female aggression squares with the image I have created of baboons using social strategies, *not* aggression, whenever possible, to settle their differences. Perhaps ironically, females, though not routinely aggressive, are likelier than males to injure each other in physical conflict. The famous early ethologist Konrad Lorenz suggested that evolution gave those with serious weapons an inhibition to use them.[2] Lacking a male's strength, size, and dagger-like canines, females can inflict more damage than most male-on-male conflict.

Now a group of more than ten baboons, including lower-ranking females, join the O's, mix it up, then leave. They have added their "two cents" while minimizing the risk of punishment from higher-ranking females. The rest of the troop watches closely as the raging baboon females race up and down the rock face. I find a good vantage point, my heart beating fast with growing anxiety for Buffalo. Nothing makes sense. Why is the O family attacking Buffalo? Female rank carries through generations, as constant as the rising sun. Or nearly. That doesn't mean a lower-ranking female never challenges a female with higher standing. But it's rare and opportunistic, usually with the support of a big male friend or in the context of a major upheaval, like the Nabo fusion. In those cases, it seldom succeeds.

Weeks go by as this heated "conversation" flares and flares again. Males try to help female friends but are driven off by their friends' foes. All of Pumphouse seems rattled and on edge. Finally, Buffalo accepts defeat and retreats to a subgroup gathered around Latour, the troop's dominant male named, yes, for my French colleague. His cohort includes Buffalo's high-ranking family. I note that Latour and his entourage have isolated themselves at the fringe of the troop. Later I see that Buffalo avoids the O family. It's unclear whether Buffalo's family, including her mother, Deborah, have also lost status.

---

Back at camp, I check my notes, consult kinship diagrams (see appendix 5), and break the situation down. Fact 1: Deborah, the highest-ranking female in Pumphouse, is aligned with Latour's subgroup of females at the edge of the troop. Fact 2: All 18 of Deborah's immediate family orbits Latour. Fact 3: Since the initial assault on Buffalo, Deborah has attempted to leave Latour's circle, but the O's chase her back. This is new behavior. Normally, Deborah's high status is a free pass to go where she wants. Fact 4: Deborah has become fixated on rejoining Pumphouse. Fact 5: Deborah's plight looks to be related to the mobbing of her daughter, Buffalo.

Deborah's behavior is baffling. Typically, she would put up a fight. I suspect she may lack the energy: She's very pregnant. Equally problematic: Deborah's adult daughters, her "posse," hang back with Latour and seem to be throwing mom under the bus. Fact 6: Pregnant Deborah finally rejoins Pumphouse once she has given birth but at a much lower rank. I see her displaced by females who had once deferred to her.

I look for reasons behind Deborah's rare and precipitous loss of rank and the emboldening of the O's. Though I'd missed the opening salvo, I sense that Deborah's family's alliance with Latour factors into the discord. They now spend their days on the edge of the group. Could it be that this spatial withdrawal created an opportunity for a revolt by the middle-ranking O's? Had it caused a shift in troop dynamics and sparked that assault on Buffalo?

Dusty was another riddle to solve. Why did she join a coup against her own family? Dusty is the middle of three sisters. That means her younger sister, Deborah, ranks above her, both in the family's reverse-age-order hierarchy and the troop's female hierarchy. Did Dusty, like the O's, see her younger, higher-ranking sister's retreat to Latour's group as an opportunity to rise in rank? Whatever her reasons, by joining the O's, Dusty rose from number two Pumphouse female to number one. Which is not to say that the O's or Dusty "plotted" an overthrow. Instead, my guess is that they saw an opening when Latour and his little group sequestered themselves at the periphery of the troop. Mulling these mysteries led me to insights perhaps better described as revelations.

---

The goal of my first study was to compare my observations of baboon behavior—specifically, the "roles" male and female baboons play within a troop—with the image of baboon society described in the "baboon model." Initially, I gave equal time to males and females, but I gradually shifted my analytical focus to males, whose roles seemed more multifaceted, dynamic, and, frankly, interesting. My first study was designed to end with an experiment in which I would trap out the large males and note changes in troop dynamics. It proved unnecessary. After 16 months with the baboons, I realized that females had a hierarchy that performed many of the functions previously ascribed to males. They, not the males, served as a stabilizing force.* The female system was based on inheritance, not confrontation, as each female's family held a certain rank that carried across generations. During 50 years of baboon watching, I have seen a few female families die off, yet *relative* family ranking remained unchanged.

Remember that a family's fixed place in the hierarchy lets even the smallest member in a high-ranking family displace a lower-ranking larger baboon. But each family also has its *own* internal hierarchy. I described this internal arrangement in chapter 6, but it bears repeating here. A family, no matter its rank, has at its center a matriarch. Her offspring are ranked according to birth order, with the oldest having the lowest rank and the youngest the highest; if sibling rivalry occurs, the matriarch will side with the younger of her offspring against an older one.† Younger progeny have the greatest potential to breed compared to their older siblings.[3] Thus, the matriarch supports that reproductive potential, creating an inverse ranking within her family. Therefore, females have two sets of rank: their family's within the troop and their standing in their own matriline.

And yet tension can flare unexpectedly, even in a stable matrilineal hierarchy. In 1997, during a session at White Rocks with Malaika, I heard eerie screams. I ran to catch up, then stopped and raised my binoculars. There was Zilla, and in her mouth was a Cape hare, the scream's source. By the time I reached her, Zilla had dismembered the hare. Baboons are usually opportunistic, not regular, predators, but are glad to get meat protein when they can.‡ This

*Males, of course, also had a hierarchy, but it was dynamic, not stable. Challenges to male rank were common, along with testing behaviors, like greetings and bluster, that could go on for days.

†An evolutionary principle called reproductive value (Vx) offers an explanation.

‡Males, the main predators, sometimes share a catch with friends, but I've also seen males use intimidation to get a portion of another animal's meal.

day, no one challenged Zilla, but two baboons, male and female, sat quietly nearby, poised to get leftovers.

Heather, Zilla's daughter—the spitting image of her mother's younger self—approached. What happened next was a shock. Heather attacked Zilla without prelude or provocation. Zilla dropped the hare, meaning that it was now fair game for onlookers. Should I watch this unusual fight or see what happens with the hare? I opted for the fight.

At first, the two females look to be "play" wrestling, but they didn't wear the "play faces" youngsters use to mean "this isn't serious." A play face has an open mouth with lips covering the teeth; exposed teeth signal aggression. Play often involves a chortle sound, a baboon's laugh. But Zilla and Heather grapple in silence. The encounter lasts 20 minutes. Heather seems the more restrained as mother and daughter fling each other back and forth to the ground. As in play, rest periods break up the action; then mother and daughter awkwardly sit together. After the third round, Zilla is visibly exhausted and breathing heavily, unlike Heather. The few times Zilla screams, a male friend rushes to her defense. He slaps at Heather, who briefly retreats; then he spots the abandoned Cape hare and rushes to claim it. Zilla's on her own.

Heather resumes her aggression. She wrestles Zilla to the ground, determined yet not vicious. Then it's over as suddenly as it began. Heather begins to groom her mother. It's a move that looks like reconciliation, but it isn't working. Zilla stays tense and tries to leave. Heather keeps pulling her back. Missing are the customary grooming lipsmacks and grunts.

I inspect Zilla through binoculars, as I have kept my distance. Surprisingly, she has only minor injuries. I see cuts on top of her tail, a small gash above her eye and another on her left hand. She's not limping and appears free of pain. I surmise that Heather meant to challenge her mother and her high rank, but not to hurt her. Perhaps she was just checking her rank against her mother's. I don't know. Soon, when Heather needs help in a dispute over food, their status returns to normal, mother over daughter.

When matriarchs die, it's an opening for rank shifts. Such shifts are rare but notable when they happen. In decades of baboon watching, I have seen some skirmishes, but only four that produced rank changes. In those cases, the outcome depended on the strength and numbers of a female's allies, especially if they were adolescent brothers and male friends. Rank changes like those of Deborah and Dusty are exceedingly rare. The tantalizing implication here is that baboon females, as well as males, are acutely aware of their standing in a troop and actually *care* about their relative rank, as Dusty seemed to. I saw this when Malaika and Soit merged as Nabo. Being part of the larger, dominant troop, Soit females weren't going to give an inch to incoming Malaika females. But a lifetime role as a high-ranking female is not easily surrendered. Several top Malaika females made repeated attempts to insert themselves above Soit females, refused to defer to high-ranking Soit females, and even tried to claim key feeding places. As we've seen, their efforts ended when Malaika males took up with Soit females. Unfriended, as it were, the Malaika females had nowhere to go but down in rank.

---

Sex and friendships link the female and male systems. As males come and go, joining many different groups over their lifetimes, they abandon family and friends and start from scratch in a new troop.* I worked a long time to figure out male dynamics. Compared to males, females seemed . . . maybe a little boring.

The Washburn / DeVore / Hall baboon model put males center stage, but a shift began in the 1980s and 1990s. Women scientists

*With each transfer, the male reinvents himself, constructing a network of friendships with females and youngsters who become "behavioral kin."

realized that female primates had been ignored so focused on them. Then inspired by the fact that, as with all mammals, there's a limit to the number of young a female can bear in her lifetime, female primates underwent an evolutionary reinterpretation.[4] It is wholly on the female baboon to gestate and feed her baby, and the energetic cost is high. Jeanne Altmann's seminal study of yellow baboons in Amboseli National Park, Kenya, described baboon females as "dual career mothers."[5] They fend for both themselves *and* their dependent baby, and thus lead lives greatly constrained by biology. Sociobiological theory posited that this constraint produced sexually selective females.[6] Whatever the theory, the reality was that a female baboon could have as many as 15 offspring at Kekopey, or up to seven after the move to Laikipia , where food was less plentiful. By contrast, a male adept at monopolizing receptive females could father seven infants with seven females in succession with a minimal energetic investment, before or after compared to females. Sarah Hrdy even claimed that male and female mammals have such divergent reproductive goals that they were like two different species.[7]

Arguments about dominance hierarchies, female or otherwise, assume that rank strongly correlates with reproductive success.[8] This makes sense for females given that higher rank brings better access to food and enhanced condition and fertility. Thus, high-ranking females would be expected to grow faster, reproduce earlier, stay healthier, and, by the end of their lives, have made more infants—the raw material of natural selection—than low-ranking females.

I had no problem with this assumption until I took a closer look at my records. After analyzing data collected over several decades, both before and after the translocation, I saw no relationship between rank and female fecundity.[9] Only recently has this discovery been confirmed by other studies that show correlations between fertility and female sociality and between effects of age and environment on the number of young a female produces.[10]

If rank has little to no bearing on a female's fertility, I had to wonder why it even mattered, at least from an evolutionary viewpoint. Perhaps a clue lay in what I'd just seen firsthand—namely, what happened when volatility upended the comfortable predictability of the female hierarchy. The entire troop shuddered. Stability vanished, certainty was gone, aggression between many baboons surged. It occurred to me that the true purpose of the stable female hierarchy may have been hiding in plain sight, visible only in its disruption. The implications were startling, and they pointed to *predictability* as a crucial product of the female hierarchy. It codifies relations between females and their families who comprise the majority of the group. To borrow language from economic theory, predictability reduces the transaction costs of baboon social relations, which otherwise can be fraught. Baboons continually monitor the social world around them because shifting relationships within the troop may bear on their future actions. In this way the female hierarchy is a hedge against disorder.

---

After several months, the Pumphouse females finally settled down. I could now easily imagine what might have happened if they hadn't. Paralysis. With ranks unknown, order would vanish. Each step past another baboon would require a new negotiation. These negotiations would consume time needed for crucial behaviors like foraging. Their very survival would be at risk.

Predictability as a tool of evolution isn't as crazy as it sounds. Brain and behavior studies in humans and monkeys show that unpredictability creates anxiety and stress. In a famous experiment, human subjects, given the choice, preferred an immediate shock to waiting for one in an indeterminate future—to avoid the stress of suspense.[11] Ambiguity stirs similar feelings of anxiety—more so, in fact, than the prospect of harm. In another human investigation, study subjects were assigned "rank" then randomly given new ranks. Neuroimaging confirmed a sharp anxiety response to the new ranks. Stress is an important measure of the way disorder and uncertainty affect individuals.[12] All this seems pretty self-evident for humans, but what about monkeys?[13]

Imagine something as basic as grooming in a baboon troop lacking an ordered female hierarchy. Normally, grooming involves a cost that limits the number of grooming partners a baboon can handle. Two studies have found that baboons faced with a surfeit of grooming partners become visibly stressed.[14] The female hierarchy can bring order to the matter of who grooms whom, reducing stress and social costs.

Both observed baboon behavior and brain experiments support my notion that the female hierarchy exists, at least in part, to codify and streamline social interactions. Thus, it minimizes transactional wear and tear that typifies baboon complexity and eases anxiety *and* stress.

If evolution ordained a reliable, predictable female dominance hierarchy for interactions, then female rank is decoupled from reproductive success. This interpretation, if correct, also solves the puzzle of why low-ranking females usually accept their rank. *All* the troop's members, regardless of rank, benefit from predictability and stability, even the lowest-ranking female. Baboons aren't the only species to have realized that order makes life more livable.

But any structure can provide order. The original baboon model claimed that it was the males and their dominance hierarchy that structured the group and provided order. I found the male system to be dynamic, ergo unstable, changing each time a male joins or leaves the troop and even when a troop's young males reach maturity. Flux in the male population—and thus in its hierarchy—makes the male system likelier to stoke social stress than relieve it.[15] Luckily for any troop, large males form the smallest contingent, making their routine eruptions less destabilizing. If females were similarly volatile, baboon troops would like be endless seething cauldrons of conflict. Thus, the female hierarchy emerges organically from a simple and deeply embedded evolutionary rule which the baboons follow, but of which they're unaware.

---

I'm intrigued by those rare times when females show that they care about rank. But if rank isn't linked to reproductive success, why would a female aspire to higher status? We've already seen one reason, in chapter 4, in the story of low-ranking Beth, who left Pumphouse to join the crop raiders. In the smaller troop, she still ranked the lowest but had fewer females above her and presumably felt less stress (and got better food). Then there were the low-ranking Soit females, determined not to let any Malaika female rise above them. The supposition is that stress increases in proportion to the number of females in a troop that rank higher. Stress has been shown to affect individual reproduction in male primates and sometimes in females.[16] Still, despite the stress of low rank, one rarely sees low-ranking females vie for higher status, even when an opportunity presents itself. I can only conclude that they accept their low rank because the female hierarchy provides the benefits of "order" for them as well.

Then there's the matter of food. Because high rank confers dominance over food, there should be a clear divergence in fertility between high- and low-ranking females when food is scarce. But my data question this assumption. The translocation provided the perfect test case. Compared to Kekopey, Chololo, being arid and drought prone, offers less food during the dry season and droughts. Yet even when food was scarcest, the fertility rates of high- and low-ranking females remained similar. The reason may be that in times of great scarcity, food is more dispersed, making it harder for high-ranking animals to monopolize what is there.

Whatever is going on, the simple evolutionary equation that rank = food = growth / reproduction is modified or overwhelmed by real-world factors that appear to be less sensitive to female rank.[17]

---

For lack of a better term, I describe baboon females as conservative. Theirs is a temperament honed by evolution over millennia of selection. Unlike males, females are risk averse, as we've seen when they balk at leaving their home range.* Conservative females seldom perturb their hierarchy. When it happens, females usually accept the results, unlike males. In fact, it seems to me that the dynamism of the male system is actually *enabled* by the orderliness of the female system. When disarray spikes in the male system—following, say, an influx of immigrant males—the stable female system just keeps on ticking, the calm eye of the storm.

Order and predictability are compelling arguments for the female hierarchy. Individuals (and groups) need a way to manage daily existence. This makes the female dominance hierarchy a *transactional evolutionary principle* rather than a genetic characteristic of individuals or their behavior. The female hierarchy is the baboon version of what humans do: "An economizing behavior in the sense that it greatly reduces the transaction costs of social interactions."[18]

We as humans seek order as a benefit, but the idea that the hierarchy evolved as an agent of order, not reproductive success, is heretical. Rarely do we even talk about how order came to be. Its existence is assumed. But how, where, when, and why did order become a feature of baboon life? I wondered if female baboons had given me a glimpse of the story behind the story.

I owe baboon females an apology. They are far from boring. Females are conservative for good evolutionary reasons. Whether the disruptive effort to rise in rank is a benefit or a cost depends

*Risk avoidance supports their biological mandate to reproduce.

on many things: family size, family standing, the ages and temperaments of relatives and friends, and opportunity. A simple choice—such as following a male friend to the edge of the group, as Deborah and her family did—can have unintended consequences. As I uncovered more layers of baboon complexity, I wondered how baboons manage. Finally, it seemed so clear: The female hierarchy forms a shield against chaos.

These thoughts gestated over many years . . . decades, actually, because I went from inside the scientific model to somewhere outside it. I blame the baboons. I merely watched. What they showed me, and what I duly recorded, refused to align with prevailing "models" and theories, even after I discarded the "baboon model." The most transgressive of my ideas were simply products of baboon behavior. I'd witnessed the disruptive impact of instability in the female hierarchy and deduced its crucial role in group order. And I'd documented a striking *lack* of correlation between a female baboon's reproductive success and her rank. It took a while to summon the nerve to stand behind these heterodox discoveries.

---

In any event, I kept thinking. Perhaps it wasn't a matter of the baboons not fitting the evolutionary framework. Maybe my interpretations had a bias, mainly my lifelong fascination with what *doesn't* fit predictions. Still, I was pleased with this reinterpretation of the baboon female hierarchy.

Chapter 11

# Mob Story (2007–2012)

**THOUGH I HAD BEGUN** to question aspects of current evolutionary theory, I was still largely in the thrall of "perfectly adapted" baboons. That meant that even as the world changed around them, baboons would react in ways that were best for them, if not always identically. But what if this wasn't always the case? What if something new came into their lives and threw everything out of whack? Something like the radical transformation I saw unfolding on this arid savanna. In a mere decade, a nonnative plant had spilled over from the little village of Doldol, then had begun an aggressive march westward, spreading beyond anything I could have imagined. It was a special cactus, and the baboons loved to eat its fruit. Competition for this new food could get fierce particularly when it was limited to just one area. A behavior I'd not seen in my decades of baboon watching emerged, and it called into question another basic tenet of evolutionary theory.

I had asked Jonah to fly me over the baboons' home ranges. This time, I'd been away from the baboons much longer than usual and needed to get my bearings. After years of misery, I'd had little choice but to seek a surgical solution to my back problem, risky though it was. Already I'd lost four inches in height. My surgery took place in San Francisco in 2006. My spine was fused with steel rods and wire cages, an operation that took the better part of two days. Rehab lasted a year. There were still limits to what I could do, but the surgery gave me my life back.

In just a year, so much had changed. Not only had *Opuntia stricta*, a prickly pear cactus, crept farther into baboon country, but so had people. We flew low around Bridal Rocks, a sleeping site for local baboons and nearby Il Polei, once just a small police post. Officers there were the ones who'd heard baboons pass by on the night of the elephants. Now it was a small village of tin-roofed kiosks and huts. Maasai settlements, bomas, had popped up on land that used to be open. Back on the ground, I would pay close attention to how the baboons were reacting to these developments.

**September 2007:** Dawn at White Rocks. I am bundled up against the biting cold of the long dry season, this time without the extra warmth of a back brace. But it feels great to be rid of that decade-long encumbrance, relatively pain-free and more clear headed. Even the baboons of Nabo seem to smile on me.* On my first morning out, they've slept at this favorite old sleeping site, not at craggy Bridal Rocks, where they sometimes sleep these days. Would I remember individual animals after such a long absence? Would they remember me? I would know the Nabo stalwarts, the mature females, but babies have been born and new males have joined. I hope the mothers' calm acceptance of me will assure their babies that I'm not to be feared.

I notice that a much smaller baboon group seems to have aligned with Nabo, though the two are clearly unmerged. I recognize them as the local control group from my fusion study, and now they're

*Recall that Nabo resulted from merged Malaika and Soit.

even smaller than before, only 14 animals, compared to Nabo's 82. We called them Musul after the name of the kopje where we often see them. What's very odd is that their large males are gone. The only adult animals are female.

The sun rises and I see that *Opuntia stricta* cactus plants now skirt the slopes and lower rocks. I count 100; a year ago there were only 10. These are newly established plants, yet to bear fruit. I trace the western and northern frontiers of the cactus invasion as I walk with the baboons over the next two weeks. I note that the main cactus patch, where the plants grow more densely, and is fruiting, is around Bridal Rocks. These plants lie outside of Nabo's usual range. I wonder if the new cacti at White Rocks were seeded by pooping baboons who'd fed on cactus fruit at Bridal Rocks.

I decide to stake out Bridal Rocks, wanting to see for myself that Nabo sleeps here as well as at White Rocks, which they'd always favored. What brought them to Bridal Rocks? Might little Musul troop be playing a part in what looks like a range expansion? I suspect that Bridal Rocks lies within Musul's traditional home range.

I am also here to help solve a mystery. The four Kenyan research assistants and two foreign interns have recently lost track of the Nabo/Musul cohort. In the evenings, no one's seen them scaling the rocks at either sleeping site, nor are they sighted in the morning, when they should be coming back down. Perhaps they've adopted yet another new kopje? They have their pick of seven in their home range. So I set my alarm for 4:30 a.m. (Jonah claims I'm a split personality, eager to rise for a baboon dawn patrol but dangerous to wake up otherwise.)

Day is yet to break and I'm glad to have a good flashlight. The night before, I was lucky enough to see Nabo/Musul ascend Bridal Rocks. Pumphouse claimed it for a while, after the elephant scare, while Nabo always preferred White Rocks. But at Bridal Rocks I can walk a narrow trail up to a small grassy plain at the top, which I do. At this hour, the baboons should still be sleeping. But I see no baboons. Then I hear their sounds coming from a rock face just below. I spot part of the troop making its way off the kopje, despite the lingering darkness. Those left behind play, greet babies, or just hunker on the rocks. A few rest in a fig tree on the rocky edge. I wait for what happens next.

In the dim light, the stragglers take off toward the thicket following Nabo, seeking to catch up with the rest. All but two, that is. An adult female and a juvenile remain. On my way down to catch up with Nabo, I pass a group of unfamiliar baboons; they eye me nervously. I figure they're part of a large local troop that often sleeps here.* I see Nabo enter a thicket a quarter mile away, heedless of the two left behind. Finally aware of their predicament, the two frantically "wahoo," meaning, *We are here, where are you?* But there's no response. Their troop has already disappeared into the thicket.

Nabo may have deserted the hapless duo, but the local troop is very aware of their presence, and is displeased. In a sudden surge, they mob the two Nabo baboons, clearly meaning, *Go away, you're not one of us*. The mobbing starts with low-intensity pant grunts and quickly escalates to screams and louder pant grunts—unique mobbing sounds. The harassed animals get the message. They rush down the rocks and head off in search of Nabo.

---

Today I solved the mystery. Nabo/Musul had indeed slept at Bridal Rocks. But as my predawn foray revealed, the troop had adopted an abnormally early departure time. Once down the rocks, they quickly vanished into the wait-a-bit acacia thicket below. By 6:30 a.m., when the research team shows up, they're long gone. But they're not always sleeping here, so the confusion is understandable.

*Our annual census told us that there were at least five discrete local troops sharing the combined Nabo/Pumphouse ranges, with at least 500 members total. As Pumphouse's range grew, additional local troops were identified.

As for their elusiveness in the evenings, another discovery soon came. When they slept at Bridal Rocks, Nabo/Musul had taken to lurking, hidden in a nearby thicket, waiting to see if the large local group would show up. If it did, Nabo/Musul waited until nightfall, then emerged and found their own sleeping rocks on the kopje. By this time, the research team had already returned to camp.* The troop had become fast moving and adept at concealment in the thickets blanketing this part of Mukogodo.† The mobbing I'd witnessed of the two stragglers now looked like strong evidence of what Nabo/Musul were taking such pains to avoid.

While I love being with the baboons, I was not keen on this area—aside from the surly local troop, which we decided to call Sisal. I'd begun to think that wherever the baboons and I went, people were sure to follow. Here, the growing density of Maasai bomas meant more people, livestock, and dogs. Back when Pumphouse slept at Bridal Rocks in the 1980s and early 1990s, I disliked it for a different reason. The invading plant then, sansevieria, was indigenous, the spiky, wild version of "mother-in-law's tongue," often seen in office building lobbies. With its water-dense base, it had become an important part of the baboon diet. Sansevieria likes disturbed areas, and humans and their grazing livestock were taking a toll on soil quality. Tracking Pumphouse here had me plowing through dense sansevieria stands, like seas of little knives. Both skin and trousers suffered greatly.

Two decades later, things had gotten worse. *Opuntia stricta* was increasingly ubiquitous. Thorns on the cactus pads were easy enough to remove from pads or clothes, but dealing with the

*For safety reasons, all members of the research team were required to leave the field before darkness.

†Chololo is a ranch on the eastern part of Laikipia. The communal land to the east of Chololo was called the Ndorobo reserve but later Mukogodo. This covered from Il Polei to the Mukogodo Forest above the little village of Doldol. The terms change as the times and usage change.

cactus fruit was a nightmare. Little hooked hairs called glochids coated each fruit. Even glancing contact with the plant invited a glochid fusillade, mostly aimed at my backside.

Over the next several weeks, clarity replaced confusion. The mobbing of the two stragglers was indeed just the tip of the iceberg. All of Nabo/Musul risked mobbing if they weren't off the rocks before dawn. The early departures and late returns were tactical avoidance. And when Sisal wasn't pushing them from behind, people were chasing them from the bomas if they ventured east to forage. The bomas held baboon temptations, but of a different sort from Kekopey's planted fields. The pastoralists had food stores, as well as chickens and sometimes baby goats that became easy and tempting baboon prey. I had pointedly asked the local folk to do what they could to frighten the baboons and protect their property. They complied by throwing stones and chasing the baboons that ventured near. When we first moved here, Maasai herders were seasonal residents. But now the Maasai were settling down and grazing their herds in place. Though less densely settled than Kekopey, Mukogodo was giving me an uncomfortable whiff of déjà vu. I had to wonder how the baboons felt about their new way of life in such an altered landscape. Clearer to me was what had attracted them to the Bridal area.

---

After trailing Nabo/Musul for several weeks, I finally understood why the baboons kept returning to this dicey location. They sought fruit from the mature *stricta* cactus fruiting only in the area around Bridal Rocks at the westward front of the invasion. Unlike some prickly pear found in the American Southwest, *Opuntia stricta* is not a desert plant; it is native to the temperate rocky southeast coast of the United States.[1] From there, it spread to the Caribbean islands and parts of South America.[2] *Stricta*, just one of 200 *Opuntia* species, fruits prolifically and invades aggressively, via fruit seeds and fallen pads that take root. Early explorers brought the plant to Europe, South Asia, Australia, and various regions of Africa.[3] Initially welcomed, *stricta* is now unwanted, except by baboons, other wildlife, and livestock. The double rainfall in Mukogodo supercharged proliferation by doubling fruit production and growth.

In the 1950s, before invasive species were a conservation concern, the British colonial government introduced several varieties of *Opuntia* to the little administrative village of Doldol. This hamlet happened to be adjacent to the range annexed by Pumphouse in 2000, after their eastward shift. Colonial officers cultivated *Opuntia* as "living fences" and decorative additions to their small gardens. Four distinct *Opuntia* species cropped up on the savanna, but only two were of interest to the baboons. I came to know them well over 15 years of study. Initially, I named each species by number, not scientific name, unknown to me at the time.

The first species, opuntia 1 or *Opuntia vulgaris* (common prickly pear), was familiar from Kekopey as a favorite dry season food. Baboons ate the pads after biting off the thorns; the few fruits following the rains were a special treat. Baboons ingeniously learned to knock the fruits from the cactus without touching the nasty little hairs, which they then removed by rolling the fruit in dirt. The fruit's goodness is laid bare by breaking the fruit open or peeling it to get at the insides.

It wasn't until the late 1990s that opuntia 2, *Opuntia stricta*, appeared on our radar.[4] *Stricta* is very different from *vulgaris*. The pads are rounder and less succulent, and its branches grow horizontally rather than vertically. Large yellow thorns on a young plant's lower pads thin out as the plant grows taller than goat-browsing height. The result is that the upper pads of mature plants are pricker-free. Crucially, *stricta* is a strong producer of fruits, as many as 10 per pad, with production spanning the seasons in a normal rainfall year. Baboons aim for the fruit and rarely bother to harvest *stricta* pads.* The plant, left intact, spreads all the more readily. To understand the changes I was seeing in baboon behavior, I needed to narrow my focus to *Opuntia stricta*, since its fruit was now central to the baboon diet and had begun to shape their lives.

Compared to native baboon foods, cactus fruit offers a large, high-quality package of nutrition. The main appeal is water and sugar, but these fruits also contain vitamins, minerals, and antioxidants. The average cactus fruit weighs about two ounces, of which 80–90% is water. The 35 energy calories in each fruit come mostly from sugars.[5] Like the baboons, I (and other humans) enjoyed the fruit's sweetness and juiciness. Because of its dense nutritional value, *stricta* fruit's appeal is similar to that of field crops. But the fruit has a nutritional advantage over field crops: Its water content is a valuable source of hydration, especially in this arid ecosystem. So it began to make sense that the Nabo baboons were willing to pay a steep price for access to the only fruiting cactus patches in the area. Just how steep a price, however, proved to be shocking—and a seeming violation of what baboons had evolved to do.

Another decisive element of this story is Musul, the female-led troop now aligned with Nabo. Musul had disappeared a few months after the Nabo fusion, and we had lost track of them. Now their numbers are even smaller. These females had what I call ecological wisdom. Their traditional home range was large, I guessed, encompassing Bridal Rocks, White Rocks, and beyond. Unlike Nabo, they knew of the *stricta* fruit at Bridal Rocks. Fruit was what they wanted to eat, and Bridal Rocks was where they wanted to be. Musul was the first to "tell" me that *stricta* was a behavioral game

*But in the absence of fruits, they started eating cactus pads in the recent drought.

changer for baboons. As a very small troop, Musul was ill-equipped to compete with the many baboons of Sisal for cactus fruit. Attached to Nabo, they'd have a chance.

Musul had taken to hanging out near Nabo at White Rocks, likely seeking the advantages of proximity to a larger troop, as Malaika had with Soit. As I began paying closer attention, I noticed distinctive characteristics shared by Musul females: long faces, a boxy body type, and "wild" eyes. The young females were fluffy, but the older ones looked a bit rough, as if battle scarred. We named the large females Brenda, Chichi, Quicken, and Jen, and we marveled at their fearlessness. They marched confidently into thickets of dense wait-a-bit acacias as ever-cautious Nabo approached nervously, wahooed false alarms, and finally edged away. I had to wonder if these Musul females, in the absence of males, had compensated by becoming uncommonly bold.

Musul wasn't meekly shadowing Nabo the way Malaika hovered near Soit before their fusion. We observed them making strenuous efforts to lead Nabo from their White Rocks sleeping site to *stricta*-rich Bridal Rocks. At first, Nabo refused to breach their border, especially with a troop they didn't know well. Musul went to work cultivating social connections with several Nabo baboons and duly began to gain influence. But the largest contingent of Nabo females still balked at following Musul to the Bridal Rocks *stricta* hotbed. A tug of war ensued, and cracks began to appear in the resistance. In 2007, we saw negotiations result in a Nabo subgroup following Musul to Bridal Rocks for cactus fruit and overnights; the rest of Nabo stayed at White Rocks. As months passed, I wondered if Musul would recruit yet more Nabo animals, but instead the defectors eventually rejoined Nabo.

Then, in 2008, a remarkable thing happened. Musul's relentless campaign finally paid off: Nabo, the entire troop, finally left the security of White Rocks and followed the Musul females to Bridal

Rocks. By now, both troops knew that the large, hostile, and intensely proprietary local baboon troop Sisal would be there. They would not be welcomed.

Mobbing is the strongest expression of a baboon group's disapproval, used rarely and usually successfully. Adult females typically instigate a mobbing, but others join. Mobbing enforces discipline within a troop or sends a warning to another troop. Intentions are very clear when a crush of angry baboons hurls itself forward, pant grunting, screaming, and making scary mobbing sounds. Thus, males are taught not to frighten babies and other troops are told, *You've come too close.* Over three decades of baboon watching, I'd never seen mobbing *not* work as intended.

Now Nabo, at Musul's insistence, found itself ensconced at Bridal Rocks, away from home and in the core area of a range dominated by large, unfriendly Sisal. The mobbings, once sporadic, then began in earnest. Stunningly different from anything I'd seen, ever, was the response of Nabo/Musul to Sisal's aggression. The mobbed baboons certainly would have "gotten the message," but they didn't heed it. They didn't vacate Bridal Rocks or the surrounding area, as you'd expect. They simply took the mobbings, day after day, sometimes several times a day. They were breaking two key rules. Baboons do not leave their home range, especially when it puts them in the occupied core area of another baboon troop. And baboons do not keep returning to a place where another troop has mobbed them. Musul's influence partially explained the first transgression, but not the second.

At first Nabo/Musul reacted normally to each mobbing. They ran away. But once the dust settled, the determined Musul females led Nabo back into the danger zone. Mobbings happened on top of Bridal Rocks, below Bridal Rocks, and away from Bridal Rocks, where the animals now competed for the cactus fruit craved by Musul. Unbelievably, the mobbing cycle continued for more than a year. Nabo/Musul, now nearly merged, occasionally caught a break by sleeping elsewhere, but then came back for more. I could only make partial sense of what was happening.[6]

---

I have stationed myself at Bridal Rocks as day breaks. Nabo/Musul are on high alert, perched at the very top of Bridal Rocks, above the west-facing ledges where they slept. They're watching for Sisal, who slept nearby but on better ledges. Normally, Sisal would start its day on the grassy area atop the kopje. But when Sisal baboons don't materialize, Nabo/Musul decides to leave the rocks. I follow as they sneak through the thicket below. My guess is that they're trying to beat Sisal to the densest cactus patch, but they're out of luck. Sisal is already there and sees them emerge into an open area between the thicket and the cactus. In an instant, Sisal masses and makes a straight-line dash toward the intruders.

The mobbing that follows is routine by now, but for me still heart stopping. The attackers are mostly females, subadults, and juveniles, all screaming, pant grunting, and making those unsettling mobbing sounds. Nabo/Musul quickly retreats, dramatically displaced from the cactus. I follow them to another cactus patch northeast of Bridal, but Sisal catches up. This time intense mobbing isn't needed. Nabo/Musul run away and forage at a safe distance over the next hour, seeking insects under rocks, stem bases of grasses, and dried acacia seeds. This puts Nabo/Musul farther east of Bridal Rocks but closer to a more distant cactus cluster. The only problem is that getting there requires traversing a gauntlet of bomas. It's later in the morning now. People are up and about and do as we have asked, shout and throw stones. Luckily, the family dogs bark but don't give chase. It's added stress for nothing. The baboons want cactus fruit, not whatever might be at the bomas. They hurry past the people, reach the cactus, and hastily harvest fruit, as if expecting yet another mobbing any minute.

So why did Nabo finally give in to Musul's pressure? Though cactus fruit is a baboon favorite, other foods were available. Nabo wasn't starving. Regardless, they kept returning to Bridal Rocks to sleep and kept seeking the cactus fruit dominated by Sisal—only to be mobbed again and again.

In 2008, two years after Musul first aligned with Nabo, the two troops met the criteria of a fused group. We gave them a new name, Namu, and began to notice that these animals were losing condition. Sisal looked to be in fine health, no doubt due to their unimpeded access to cactus fruit. Yet the Namu baboons, whose more diverse diet included some cactus fruit, were visibly in decline. I could only guess that days filled with frantic evasion—from Sisal, people, and dogs—were taking a toll. Conflict over *stricta* fruit persisted. Had Nabo opted to return to their traditional home range, none of this would be happening. It didn't fit with what I knew about baboons, and in fact seemed incomprehensible.

---

Watching the Nabo and Musul baboons lose their healthy glow was bad enough; then they began to die. Most vulnerable were the babies, the first to go. During the mobbing time, none of the 13 babies born survived past two years. Three died because their mothers died; some had looked healthy and normal before simply disappearing; others vanished after showing signs of illness. I use these verbs advisedly. One rarely sees a baboon die or comes upon baboon remains. A baboon is present until it isn't.

Then adult females began dying. This too happened in ways I'd never seen. Tuppence, a Nabo female, was the first. Having slept at Bridal Rocks, the baboons were descending, doing their best, as usual, to avoid Sisal. Tuppence was somewhere in the middle as the troop headed toward a cactus stand. Suddenly, she lay down on her stomach, her breath labored. Her infant jumped off her back to explore and play, as usual. No one suspected this was the end. We continued on without her, thinking she'd rouse herself and catch up. At the end of the day, I realized Tuppence was gone. We never found her body or her infant.

The next to go was Rebecca, a descendant of Robin, who'd had a special place in my heart. In 1973, Robin was the little infant who gently touched my back as I sat collecting data on her mother, shy Naomi. Adult Rebecca had a pointy nose, her mother's unique body shape, and a short pencil tail. Like Tuppence, Rebecca had come down from Bridal Rocks with the others, then began gasping for air. She made her way to a small acacia tree and sat beneath it, her infant at her side. After the experience with Tuppence, I decided to stay with her. At midday she seemed stable, so I left to observe the rest of the troop. I assumed she'd be there when the troop returned from foraging. But she was missing when we counted heads in the evening, and by then it was too dark to look for her. The next morning, we found her body near where I'd left her. That in itself was unusual. Her infant survived with help from older siblings and a male friend.

With Rebecca, we had a body that could be autopsied. Clearly, we needed to know what was killing these baboons. The pathology report from the Institute of Primate Research in Nairobi, the same institute that helped me capture the baboons for translocation, came back. Rebecca's lungs showed massive calcification and had simply ceased to function. We were incredulous. She'd seemed normal the day before. How could lungs suddenly fail with no earlier signs of distress?

The next to die was a Musul female, Brenda. The pattern was the same: Brenda came down from the sleeping rocks with the troop, had trouble breathing, and collapsed, dead. By now, field staff carried black plastic "collecting" bags and rubber gloves. Brenda's postmortem showed that she had also died of asphyxiation, her lungs hardened. She too had seemed fine the day before.

In discussions with vets and animal handlers, I learned about "shipping fever." Animals that appear strong and healthy have been known to die suddenly after being captured, handled, or shipped, even when humane measures were taken to minimize trauma. The root cause was thought to be stress.

The loss of Wiggle hit me especially hard. He was a favorite, a handsome male with a beautiful coat of hair, black hands, and "kind" eyebrows. I'd watched him since his birth in Soit. Wiggle had a following, a devoted group of youngsters who relied on his protection. He even became surrogate mother to an orphaned infant. But being a strong defender of friends, as many males are, put him at risk. This day, Nabo/Musul had descended from Bridal Rocks around the same time as Sisal. Sisal then launched a particularly intense mobbing. Contact aggression is rare, but Wiggle was badly wounded as he defended a female friend from Sisal's aggression. Afterward, his little subgroup of female friends and their families disappeared for a few days. Shortly after they returned, Wiggle died. His symptoms reminded me of those I'd seen in animals who'd been injured in falls from rocks or cliffs: a shortness of breath and lethargy. Yet in light of the other untimely deaths, I had to wonder if mobbing-related stress had played a part.

Over several years, five adults and most of the infants died because their troop chose to live where they were unwelcome. It seemed all cost with no benefit. Why were Nabo and Musul willing to accept such abuse—even invite it? Mobbing is meant not only to punish but to instruct. But Nabo and Musul, and later, as fused Namu, refused to learn, or comply, very much to their detriment. They kept sleeping where their tormentors slept and they kept seeking cactus fruit their tormentors wanted for themselves, even when other food was available.

A strange thought occurred to me and would not let go. Could it be, I wondered, that these adaptable and smart baboons were making a mistake?

1. Hall, a large male

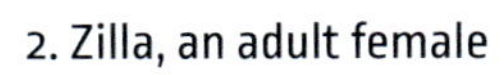

2. Zilla, an adult female

3. A male showing his hardware

4. A female doing an eyelid threat

5. A family

6. Elf's orphans

7. Male and female friends

8. A male with an infant buffer

9. Friends grooming their male

10. A young infant

11. The translocation team

12. At sunrise

13. The Kenyan team

14. Fresh grass

15. *Opuntia stricta* fruit

16. Troop movement

17. Carissa, Guy, Shirley, and the baboons

18. Young baboons playing

19. Jonah and the baboons

20. Baboons in hats

21. Twala women

22. The baboons' favorite sleeping site

Chapter 12

# Finding Meaning in a Mistake (2012–2015)

**PERHAPS THE HARDEST PART** of fieldwork is bearing witness to harm. I talk about the challenges of managing your feelings and maintaining scientific objectivity in chapter 17. It was something I struggled with from early days in the field and still do, even now. I am human and can't help but worry about the baboons I have spent so much time with and gotten to know so well.

The deaths of Tuppence, Rebecca, Brenda, and Wiggle were, perhaps strangely, a little less agonizing than some I'd seen. It might have been because no human intervention could have saved them, though I'd have liked to have sat them down and explained why daily mobbing was a bad idea. They had died suddenly, with no prior signs of distress. Also, during the two years it took for Nabo and Musul to fuse, the baboons had chosen a path that didn't make sense to me. Whatever was happening needed to run its course if I would have any chance of understanding it. Still, I had a sense of rising anxiety. Would these deaths continue and put the troop's very survival at risk? Could it be that I'd caught a glimpse of extinction in its earliest stages? It was a possibility that made me wonder if the baboons had defied the most basic evolutionary principle: survival.

**September 2012**: Dawn at Bridal Rocks. The view from the north side of the rocks is expansive. On a clear day like this, I can see the Mukogodo Forest to the east, Mount Kenya's glistening snowcap to the south, and White Rocks to the north. I am with Namu, the troop formed when Nabo and Musul fused in 2008. Namu now sleeps on the "good" side of Bridal Rocks, where the kopje is steep and less accessible to leopards. Sleeping there as well is Sisal, the large local troop that relentlessly mobbed them. Four years on, there's still a clear separation between the two. At the first glimmer of sunlight, Namu quickly moves to the northern end of the rocks. There, farther from Sisal, they can greet the day in peace, sunning and resting.

Though it cost them dearly in loss of life, Namu's persistence may be paying off. An uneasy truce is in place. The Namu baboons still get displaced by Sisal, but mobbing is rare at Bridal Rocks. Often, Namu is first to leave the rocks and keeps moving as the larger group follows. Sometimes, Sisal gives chase, but then stops once Namu reaches the far side of the gully, moving into their old home range—home free, so to speak. Namu will sometimes spend the night at White Rocks, but more typically ends their day at Bridal Rocks. Rains have come and foods are plentiful, including cactus fruit. In the evenings, they still hang back, waiting for the larger group to ascend and settle before making the climb themselves.

I'm cautiously encouraged by Namu's progress. Most newborns are surviving infancy, and the death rate is only slightly above normal. Still, it would be a stretch to say the baboons of Namu are models of robust health. They are thinner than the local baboons, their hair has less sheen, and the females produce fewer babies than those in Sisal. Their default demeanor is vigilance, not what you see in a group of baboons who lead lives free of stressors.

I struggled with these facts because the baboons had convinced me that they are smart in every way. Yet it seemed not so smart to be sharing Bridal Rocks and competing for cactus fruit with unfriendly local baboons, year after year. A tacit cost-benefit calculation had always seemed to underpin the baboon behaviors I had seen. Over time, most animal groups seem to figure out what's good for them and what should be avoided. It's a form of rationality. But it looked like Namu had chosen a way of life that cost them dearly and had few, if any, benefits. I'd seen how the cost-benefit analysis had paid off for the crop raiders. Better nutrition had outweighed the manageable risk of human hostilities. Namu had chosen to sleep at Bridal Rocks, the site closest to the most productive cactus patches, a sensible calculation. But Sisal limited their access to the fruit and maximized their stress with mobbings. Not sensible. Up to now, the baboons I had studied had shown great acumen in solving the problems of survival, even faced with novel challenges such as their translocation. Behaviors like Namu's now might have doomed that effort.

There had to be an explanation, but the only tool I had was my own deductive reasoning. Experiments with monkeys had demonstrated their ability to correlate certain behaviors with certain outcomes, even in the abstract.[1] You would think this level of cognition could enable a baboon to correlate constant mobbing with the sudden deaths of troopmates, but it didn't. Perhaps the baboons failed to make the connection and count those costs. Yet I could, and I did.

Could it be that baboons only cared about the deaths of someone close, a family member or friend? I had seen mother baboons "grieve" a dead baby, carrying it around for days, reluctant to let it go even as its little body decomposed. Though it was unusual to lose most of the babies born in one year in one troop, infant death is not in itself so rare. The possible causes are many. Perhaps Namu hadn't noticed how many babies they'd lost because the deaths occurred sequentially, not all at once. It might have seemed normal to them, simply the "circle of life." If they'd put two and two together, I'd like to think they'd have ceded Bridal Rocks to Sisal and slept elsewhere. This is what evolutionary theory would have predicted. That meant Namu failed to equate their choice to stay at Bridal Rocks with harms caused by Sisal. Lest we forget, it was all about access to cactus fruit.

Even if the baboons couldn't count the costs—the loss of condition and lives—which are, after all, abstractions, they certainly experienced the mobbing. So the most pressing question was, why did Namu dig in at Bridal Rocks, even though they knew it meant endless mobbing? Mobbing aggression happens in real time, and the message is clear: *Don't do that again*. I had seen countless mobbings, and never had I seen them *not* achieve the desired deterrence. At Bridal Rocks, Sisal was telling Namu, *Just go away*. But Namu kept coming back, despite the clear message. Nor were they defiant. They'd still run away from a mobbing. And yet, by 2008, even knowing the abuse that awaited them, they chose to sleep at Bridal Rocks most of the time, same as Sisal. And they regularly got mobbed. Finally, in 2012, they began to spend more nights at White Rocks, but this had been an option all along. Namu could have slept at White Rocks, avoided aggression, and trekked the fairly short distance to the Bridal Rocks cactus patch first thing in the morning.

After a good deal of puzzlement and mulling, I reached an unnerving conclusion. The baboons were making a *mistake*. Again and again they blundered back to Bridal Rocks and slept near the local baboons, thinking this was the best way to get cactus fruit the next day. They persisted despite being mobbed, despite the deaths of adults and infants, and despite their worsening condition. If they had somehow calculated the costs and benefits, they'd gotten it wrong. If they had decided this was the best way to get cactus fruit, they got it wrong. People get things wrong all the time. Why wouldn't baboons as well?

My struggle with this idea was rooted in principles I'd accepted as fact. The first was evolutionary theory, which assures us that wild animals are well adapted. Watching baboons convinced me that this evolutionary proposition was correct, at least for "smart" primates. Evolution had honed baboon ways over millennia, and time and again the baboons showed me how smart they were and how well they manage, no matter the circumstance, in real time.

On the other hand, why shouldn't baboons make mistakes? I had to interrogate myself about that. Was I really convinced that evolution had designed animals to be perfectly adapted to their environments? I never really thought to challenge this idea, nor had I sought evidence to the contrary. I had struggled against the scientific tide by documenting how smart baboons were, both socially and ecologically. Baboons had shown themselves to be creatures able to quickly read and react to changes in their surroundings, some of which had been profound. I saw this when people and crops came to Kekopey, and again when I moved the baboons to Chololo. I had documented their decision-making processes around troop movements and range shifts, and I had shown that they have options and don't always agree with each other. Sometimes baboons will choose to leave a group rather than accept the consensus, as when the crop raiders split off to become the Bad Guys. If the Namu baboons had made a mistake, did that mean the Pumphouse *non*-raiders were also mistaken? Recall that when we assessed the animals before the translocation, the raiders were in better condition from eating field crops.

Was I falling down a rabbit hole? I assumed that smart and adaptable baboons had good odds for survival in most any situation, and indeed the species had done very well. And yet there are no evolutionary guarantees. The uncovered history of life on earth tells us that more than 98% of all species have gone extinct.[2] Were the Namu baboons on a path to extinction? After all, most extinctions begin with one death and one group at a time. We think of extinction as a function of natural cataclysms (dinosaurs), man's depredation of nature (orangutans, almost), and extermination (passenger pigeons).[3] Perhaps it also includes mistakes.

I finally surrendered to this line of reasoning but had little success convincing others, including Jonah. He emphatically rejected the notion that baboons are capable of making mistakes. I owe a lot to Jonah. He fashioned me into a savanna ecologist, and he'd stood by me in crisis after crisis. But here he drew the line, and he used my findings against me! Jonah argued that since I had convinced him that baboons were smart, the Namu baboons must have made the right decision, despite the costs. Perhaps, he suggested, I was looking at the wrong time interval; perhaps I hadn't watched "long enough." We argued back and forth, and I kept watching.

**September 2013**: Maybe Jonah was also right. Now, nearly six years after Nabo and Musul fused as Namu, I see stronger hints of a good outcome. A *few* young Namu females, those who grew up eating cactus fruit, are reaching menarche earlier and have shorter interbirth intervals. Touché, Jonah.

But Namu infants and adults are *still* twice as likely to die as those in Pumphouse, who lead comparatively easy lives many miles to the east. An improved birth rate doesn't matter if the infants don't survive. The Pumphouse range lies near the origin of the cactus invasion, and Pumphouse never had to fight for its fruit. Namu's days still revolve around avoiding harassment near Bridal Rocks, and its cactus fruit obsession continues to take a toll. Older males and females, not just mothers with infants, are still in poor condition.

Later this year, Namu finally begins to break free from Bridal Rocks. Back in their old home range, they focus on "rainy season" foods: flushes of green grass and herbs and tree foods like acacia flowers, pods, and seeds. While smaller in size, the pods and seeds contain more protein and fat than cactus fruit, nutrients critical to growth. Perhaps Namu's return to White Rocks was inevitable. Overgrazing, charcoal making, and new settlements have led to declines in tree cover and soil quality around Bridal Rocks. Still, when the rains fail in 2014, Namu returns to Bridal Rocks, seeking cactus fruit.

I predicted that *Opuntia stricta* would continue its invasive march, a blessing for the baboons. I was right, and as I'd hoped, the growing ubiquity of cactus wound up solving Namu's problems, and some of mine.

**September 2015**: Early morning, back at White Rocks. *Opuntia stricta* now speckles the landscape farther west and north of Namu's original range. Once again, Namu's preferred sleeping site is White Rocks, where they have a key advantage: They are dominant to other groups that might sleep here.

Namu's females are reaching menarche earlier and having more babies, closer together. Namu's physical condition is decisively better than it was during those desperate years at Bridal Rocks. But it's yet to match the robust condition of the Pumphouse baboons, with their access over the past 10 years to uncontested *stricta* fruit.

Who was correct, Jonah or me? I think Namu made a mistake by most any measure. They'd subjected themselves to chronic mobbing and the result was poor condition, infant and adult deaths, and a distorted and stressful "activity budget." Their waking hours were spent dodging the local baboon troop or being mobbed by it. They did, however, achieve a Pyrrhic victory of sorts. They'd managed to carve out a new section of home range around Bridal Rocks, in another group's domain. So perhaps Jonah was also correct. Yet eight years on, Namu had abandoned Bridal Rocks and returned to White Rocks, where young cactus had continued to spread and grow into fruit-bearing maturity. Now Namu had their own good supply of *stricta* fruit. But the biggest benefit may have been the absence of a hostile rival troop. How would they have fared if the cactus had stopped spreading? Probably not so well.

---

Scientists studying primates, and many other mammals in many different habitats, now use a uniform set of data collection methods. For example, an investigator selects in advance individual animals—"sample" animals—to watch. Sample animals are chosen systematically in order to minimize bias and address the study questions. A set time is spent observing the sample animal using proven techniques such as scans or focal follows.* Methods are precise and as close to objective as one could hope. That means that I can glean information from a scientist I've never met about baboons I've never seen. But it is a large step from data to interpretation, particularly because the baboons I studied kept changing, as did the places where they lived. Each shift from the norm required me to reassess my earlier conclusions. I needed to keep rethinking baboons, and also the way we study them.

Despite progress in data collection there is still no agreement on interpretation (see chapter 18). Careful description helps, or at least it did in my early days of primate studies. Hiro Takasaki, a Japanese primate scientist, once said: "Good description will last forever, while theory is ephemeral."[4] Better understanding came from better quantification and more sophisticated statistical methods. Now, advanced computer models can take small bits of data and simulate the likelihood of various outcomes. But even with these improvements, interpretation can be slippery, as I discovered at each step of my baboon voyage. Interpretation requires clear

*A focal follow involves following an individual animal for a predetermined amount of time, regardless of what the animal is doing (e.g., grooming, nursing an infant, or even sleeping).

reasoning backed by data, and educated guesses are best couched in deep experience with the study subjects.

Yet scientists are still stuck with their "human" brains, which tend to see the world in certain ways. Evolutionary arguments often use what is called the "adaptationist paradigm."[5] I had been socialized into the implicit adaptationist assumption that animals in nature are exquisitely equipped to lead successful lives in the place they inhabit. Hewing to this paradigm forced me to resist the idea that baboons could make mistakes, let alone consider the implications of such an idea.

And yet . . . despite hundreds of thousands of hours of observation, and given the fact that half of wild primate species are currently endangered, we have no *behavioral* evidence for the *process of extinction*. While this oversight reflects scientific prejudice, it also offers an enticing opportunity for new investigations. As groups get smaller, they no longer represent a robust "sample," so the study ends before the end happens. What if we kept going?

Social, cultural, and historical milieus also influence interpretations. Today, the study of animals in the Human Age, what scientists call the Anthropocene, requires new categories of data collection and more sophisticated interpretations of time.[6] Ecologies of the distant *past* changed imperceptibly, for both baboons and humans. Now most, if not all, ecologies are *human dominated*—and if not directly, then indirectly via the human activities behind climate change, habitat loss, and fragmentation as well as hunting and pollution.[7] Rapid habitat transformation, coupled with longer scientific studies, make it likelier that the demise of a group or a population will be witnessed. But it won't if scientists choose not to study this facet of evolution. In any case, abrupt changes occurring in Namu's ecological context might well have produced behavioral aberrations. This likelihood would seem to strengthen my interpretation of Namu's self-destructive behavior as a "mistake" made as they adjusted to their altered surroundings. Extinction must have a starting point, and in some cases it may look like Namu's mistake.

---

To my delight Namu is flourishing now, with 140 individuals, down from an impressive peak of 174. Babies and their parents aren't dying, thanks to increasingly robust supplies of *stricta* cactus and its fruit, which I'd predicted. Namu avoided becoming an extinction case study, and I'm glad it did. Indeed, this troop's history highlights shortcomings in the way we apply evolutionary principles. Recall that this group started as the Bad Guy crop raiders on Kekopey, then morphed into Malaika, then Nabo, then Namu, each distinctive, each with its own dynamic story. That tells us that our observations at one point in time will not represent a group's history or its future. By now I had seen too many exceptions to believe in a perfect adaptive fit. Even the most meticulous research may not reveal, in real time, the most important consequence of a behavior, or whether a baboon or a group, given choices, is taking the actions likeliest to assure its survival.

For most of my scientific career, the measure of female success was the number of surviving young she produced. But surviving until when? That is an important decision. What if those offspring don't survive, or the progeny of the next generation? When should we stop counting? In my early studies of Pumphouse, high-ranking Peggy's family was the largest. Now all but one is gone. Today, a different high-ranking family contributes two-thirds of the natal individuals in Pumphouse. New genetic studies provide better measures of a female's contribution to the group and population. But I wonder how much this knowledge really matters, since even these sophisticated methods don't address an issue I think is critical. Namely, *how a baboon gets into the evolutionary future,* in addition to *who* gets there. It seems clear to me that an evolutionary story that fails to consider the process of "how" is incomplete.

The baboons have shown me that the power of any evolutionary argument should depend on capturing as much *natural* history as possible (see chapter 16).[8] I had seen firsthand the ways that behavioral flexibility and adaptability contribute, often crucially, to a species success. This means any evolutionary calculus should include knowledge of a group's history and its context. Even with evolutionary criteria like body condition, female reproduction, and individual survival, I would have reached different conclusions looking at data from 2008–2009, or from 2010–2012 or 2015–2017. That is the value and the beauty of long-term studies, unbound by hypothesis testing.

Evolutionary explanations tend to jump levels, from daily life to evolutionary time, often through what we call a "black box": Behavior enters, something happens, and the evolutionary results emerge (see *The Mermaid's Tale* below).[9] The black box contains the *process*, and it's mostly ignored by primate scientists and even biologists. The baboons convinced me that the real story was happening inside the box—the process—and it deserved as much attention as the outcome, if not more.

I had inherited an evolutionary framework that the baboons were happy to flout. Could it be that the framework itself needed modifying? In sorting out my thoughts, I benefited greatly from a remarkable book by two Penn State anthropologists, Kenneth Weiss and Anne Buchanan: *The Mermaid's Tale: Four Billion Years of Cooperation in the Making of Living Things*.[10] The crux of their thesis is the idea that "survival of the fittest" is the wrong metric. Instead, they argue that survival hinges on being "good enough." This concept precisely captured what I'd seen in the baboons, and it relieved me of much self-doubt. Here was an articulation of the ideas that had taken shape in my mind because I'd seen how baboons really are and what they really do, including make mistakes.

Sixty years ago, Washburn, my mentor, applied the "good enough" idea to the evolution of primate anatomy. For example, he argued that modern humans have back problems because the change in anatomy from knuckle-walking or brachiating ancestors to upright stance wasn't "the best" solution. It was instead just "a solution" created from available parts. Therefore, the human backbone wasn't created for bipedalism but was "good enough" to do the job of keeping us upright—as mine did, until it didn't. Wouldn't the same argument apply to *behavior*, an idea he had not considered?

A more "tolerant" evolutionary process would give baboons leeway to make mistakes and behave imperfectly in other ways, and yet keep going, even thrive. In the Mermaid model, "slippage," or a suboptimal fit between the individual and its environment, is possible, leaving much room for individual variation and change.[11]

The baboons had forced me to pay attention to scales of time, and this gave me perspective. I could see evolution's tolerance in action, including an allowance for slippage that occurs when an animal's ecology is altered by unnatural forces. This evolutionary interpretation of animal behavior aligned elegantly with my observations of baboons, far more so than traditional natural selection's reliance on competition between individuals for limited resources and reproductive fitness.

And it was this approach that opened my eyes to a subject so fundamental, so a priori, that little thought had been given to it beyond the obvious. That subject was the group. Despite reams of speculation about the reasons for life in a group, no one talked about how the group came to be in the first place and, once formed, what kept it together or caused it to fall apart.

Chapter 13

# Group Think (2008–2018)

SEPTEMBER 2013: Early morning with Pumphouse, or at least what's left of it. You may recall adult male Latour from the chapter about upheaval in the female hierarchy (chapter 10). He'd formed a subgroup composed of high-ranking females and their young, including Deborah. For reasons unknown, they'd stationed themselves at the edge of the group. It was this separation that seemingly incited rebellion by a group of lower-ranking females. Among repercussions of this rare insurgency was the defection from Pumphouse of Latour and his subgroup. A new fully fledged troop was born. We named it Enkai, meaning "the spirit that is all around" in the Maasai language. This new group includes several of Deborah's daughters but not Deborah herself, who stayed in Pumphouse though she lost rank. Another daughter, Beka, joins a different Pumphouse splinter group, which hovers near Enkai. But Latour will soon be moving on, as male baboons do. Several male migrants popped into Enkai, as if to try it on for size. None stayed. Maybe they agreed with me that Enkai wasn't viable with only 14 baboons.

Like Musul, Enkai finds itself with no large male for a while. This should be an attraction to male migrants, but seemingly it's not. Unfortunately, I have yet to master the art of baboon mind-reading and can only speculate about this strange male avoidance. Finally, a subadult male joins, giving the troop a total of four adolescent males and perhaps a sense of protection.

The exit of Enkai was only the tip of the iceberg. Over the next several years, Pumphouse underwent serial and substantial rejiggerings, steadily adding and losing personnel in a way I'd never seen. But as I have learned, exceptions to the rules very often yield new insights. And in this case, what I saw pointed to an entirely new idea for that most fascinating and in some ways mysterious entity: the group or, for baboons, the troop.

In 2011, Pumphouse had 59 baboons. Then Enkai left and became a troop. In 2012, another group of more than 20 animals split off, led by Olivetti, the adult son of the feisty O family. Joining Olivetti were Dusty, who replaced her sister Deborah as the top Pumphouse female, most of the O's, and part of Peggy's family. Meanwhile, the O family matriarch stayed in Pumphouse, along with Deborah and the family she started after her loss of rank. It was getting complicated.

A new pattern emerged, in a sharp deviation from the norm. Subgroups formed by single males and their female and juvenile friends were peeling away from Pumphouse. By 2015, the troop had dwindled to 50 baboons. Pumphouse was still in the normal range, while the splinter groups, of course, were smaller. With the ever-shifting numbers of troop members due to migrating males, babies being born, and the occasional "disappeared" baboon, basic arithmetic census-taking could be a challenge.

---

Hours later I'm back at camp, poring over many pages of baboon genealogy. I have taken considerable pains to chart the labyrinthine family histories of each troop we've studied since the early 1970s.

Since my role has grown more managerial over the years, and since I still teach in California, and since I'm not getting any younger, these charts are indispensable. There was a time when I could rattle off the names of any matriline without looking at their written histories, given the first letter of the matriarch's name. But our alphabetic system crumbled over time as more baboons came into the study.

By 2018, I was on the verge of throwing up my hands. Keeping track of all the Pumphouse fissions and fusions since 2012 felt almost impossible (see appendix 6), as if I'd been asked, untrained, to direct air traffic at O'Hare. People would ask how the baboons were doing, and I'd simply say "fine," sparing them the Byzantine reality. Baboons were breaking their own rules, rules I had assiduously documented over decades. Maybe I had reached the limit of my ability to understand baboons, to make sense of the behaviors I saw.

But I kept watching. Gradually, the mist cleared. I had found a way to elaborate the process of the dynamic balance of forces that bring baboons together and pull them apart. As before, the baboons taught me not to rush to conclusions. My worry that Pumphouse verged on disintegration led to one of those treasured moments of scientific insight. I finally figured it out.

---

I am now face-to-face with the major challenge of this book, conveying the incredible complexity of baboon lives without drowning you in details or boring you to death. It took me years to figure out what to do with so many actors, so many troops and subgroups, and so much backstory. It becomes *too much* to grasp, even for me at times. I imagine the reader slamming the book shut, confused or, worse, bored, and then missing the baboon insights that follow. My strategy was to present the complexity as I discovered it and hope that I could lead you there.

Like this book, this chapter has had several lives. The version before you, like most chapters, opens with baboons, then segues into the realms of observation, interpretation, context, and theory. Only this time, a major theme is social complexity. Complexity is a word apt to cause eyes to roll. But applied to baboons, it's actually a simplifier, like a prism that merges many colors into just one, white. With baboons, complexity describes not only a rich behavioral spectrum but the demands it puts on both individuals and the group. Please keep this in mind as we proceed.

Pumphouse had become an epitome of complexity on overdrive. Six independent social groups emerged over six years. Their names, aside from Pumphouse, you are under no obligation to retain: Wooro, Ngela, Enkai, OG's Troop, and Yohan's Troop. Initially, I thought this swirl of shifting troop allegiances would be most comprehensible as a play. Each act would focus on certain groups and subgroups. The first draft of this chapter had only four acts, then four became 12. It seemed unwieldy, so I thought to write an introductory summary. Jonah offered his opinion: DON'T. After a few more stabs at playwriting, I decided Jonah had a point.

---

When I watch the baboons, my process is inductive. I receive information directly from them in the form of their behaviors. But when the time comes to derive meaning from these behaviors, I start with what I know, then begin extrapolating. In this case, my starting point was negotiation. I already knew that baboons negotiate their social group (chapter 9), as after the pulse of predation that inspired Malaika and Soit to fuse. I now revisited this idea. The twelve Pumphouse fissions and fusions suggest that negotiating a new or reconfigured social group might be more commonplace when certain conditions are met. Fusions are rare. Fissions happen more regularly, and the majority of the 12 acts in my would-be drama are fissions.

This got me thinking about a baboon group's purpose. Two evolutionary theories offer competing explanations for primate "socialness." One emphasizes the benefit of protection against dangerous predators. Living in a group provides more eyes and ears—and bodies—to detect dangers and avoid being eaten.[1] The other claims the group can win in feeding competition with isolated individuals and with smaller groups.[2] I set aside other known advantages of group living—access to mates, playmates, grooming partners, and shared knowledge—because they carry less theoretical weight today.

I began by asking myself if something in their environment might have triggered such an extreme flux in the Pumphouse population. The cactus invasion stood out. Initially, *Opuntia stricta* was concentrated only in one area near Doldol, its place of origin and where Pumphouse had established a new home range. There, Pumphouse was routinely displaced by larger local baboon troops. They had to wait their turn. Then the cactus spread, dispersed by elephants, livestock, and people, as well as baboons.

Equally significant was a dramatic change in the Maasai way of life around 2000. Maasai no longer seasonally grazed their cattle across an unfenced savanna. Now they were settling down, creating permanent bomas and grazing their livestock in the same area. The results were overgrazing, irreparable harm to the land, and a tipping point for the cactus invasion. With rangeland destroyed, erosion accelerated and soil temperatures increased. This "perfect storm" created ideal conditions for *stricta* cactus to flourish and spread.[3] As *Opuntia stricta* proliferated, Pumphouse no longer faced competition with larger troops for its fruit. This released an important ecological constraint, one that allowed Pumphouse to eat delicious *stricta* fruit without suffering for it, as Nabo had to.

But the 12 acts in the Pumphouse Play had me delving more deeply into the root causes of Pumphouse's fractures and fusions. Usually, a primate field scientist doesn't encounter her "group in the making"; it already exists and persists despite births, deaths, and migrations. She only needs to know that group living, being social, confers many advantages. Period. But Pumphouse was teaching me again that a group, no matter its configuration, isn't a given. It is the outcome of a process, which means that this otherwise cohesive and important organ of adaptation can be challenged by a variety of pressures. The evidence I now have confirms what Bruno Latour and I argued many decades ago.[4]

Today, I better understand how a group is *negotiated*. I now imagined that the group was the result of a dynamic equation. On one side are forces that pull baboons together. On the other are disruptive forces—some internal, some external—that push baboons apart. A stable troop balances these forces. But when balance is lost, the group will reshape itself, splinter, merge with another, or possibly even cease to exist.

Let's first revisit the two principal forces that pull baboons together: protection against dangers, most notably predators, and a larger group's advantage when competing for limited resources. So what's going on when a baboon group loses cohesion? To find out, we need to look closely at the rich mix of social and ecological forces that can act as agents of rupture.

Most obviously, large groups are harder to coordinate (an assumption in the literature but not yet proven) and must cover more ground to find enough food (as has been shown for many primate species). Feeding competition among members increases with group size. This explains why large baboon groups split into temporary but smaller foraging groups when food is scarce, as do other types of wildlife. But as a troop's population grows, more conflict occurs, a correlation that makes sense: The more animals in a group, the greater the social complexity and the greater the potential for social friction as well. Keeping track of the dynamic social relationships around them is a mentally taxing job, even for a large-brained (if nonhuman) primate.

What I'm saying about the pros and cons of group living isn't new. What is radical are my ideas about the real-time dynamics that create a group's cohesion or cause that cohesion to fray.

Normally, we observers experience a dynamically balanced social group, but Pumphouse showed me what happens when this balance disappears.

The *stricta* cactus invasion was one important piece of the puzzle, but there were others as well. Pumphouse and her daughter troops still needed protection from the leopards that regularly killed baboons at night. We saw that the formation of a larger group protected Nabo from some predation. Yet leopards didn't target the small Pumphouse splinter troops because these offshoots found an ingenious way to protect themselves. They slept close to each other and to Pumphouse, availing themselves of many adjacent sleeping rocks.* Any of the small groups could rely on a collective of baboon troops to be on the alert, meaning that they hadn't lost the "safety in numbers" advantage so important at night. But the next morning, they would depart as small groups, leaving the social complexity of a larger group behind.

A stable social group has found a "sweet spot" in which all elements of the equation are roughly equal. But, as we've seen, any number of factors can tip a troop into imbalance, and Pumphouse had become the poster child for group disarray. Simply put, more baboons means more conflict, and social conflict disorganizes relationships and social structure.

---

Recent brain experiments with humans and nonhuman primates bolster my arguments. Let's start with the long-simmering rivalry between two theoretical explanations for the large primate brain. The winning side claims that social complexity grew our brains, citing the correlation between brain size and the size of a group

*I speculated that Namu did not fission once the *stricta* cactus became ubiquitous because they didn't have sleeping sites that allowed both separation and proximity. This obliged them to remain intact for safety reasons.

across all primate species.[5] Humans, of course, reside at the distant top of the heap. The side now considered wrong attributed brain growth to a species' response to their specific ecology.[6] Later I will argue that this is a false dichotomy.

New data on how the brain works support the notion that baboon social complexity, which increases with group size, comes with cognitive costs. There are so many things to consider for each interaction embedded in socializing, foraging, troop movement, and many other activities. We know that the frontal cortex of the brain is the executive center for the social component of primate behavior, among other cognitive functions.[7] Comparisons across different primate species show that those living in bigger groups have larger frontal cortices.[8] As the frontal cortex is tasked with decision-making, memory formation, and, in particular, social behavior, the result is an energy-consuming "cognitive load." Research shows that when humans experience a high cognitive load, other frontal lobe functions suffer.[9]

Interestingly, human studies have also shown that heavier cognitive loads make individuals less "prosocial"—less inclined to be "nice" to others.[10] A comparison of different monkeys and apes found that the few primate species in "fission-fusion" societies, such as chimpanzees and spider monkeys, have inhibitory circuits that enhance social behavior. Other primate species living in permanent groups, such as capuchins and macaques, lack this mechanism.[11]

Most brain studies of cognitive load use human subjects and assess their sociality by quantifying the subjects' social networks, including social media, phone records, and text messages. In an equivalent study of rhesus monkeys—considered the baboons of India—animals were divided into social groups of various sizes. After several months, the researchers found that monkeys placed in the larger groups had experienced growth in the frontal areas of their cortices.[12] That meant the larger social networks increased the frontal areas of their brain.

Too much complexity, however, overloads the brain and leads to bad decisions and memory malfunctions. These brain studies suggest that baboons *should* choose to live in smaller groups to minimize social complexity, if they can. Living in a larger group would naturally, and not helpfully, inflate the "cognitive load" of individual baboons who, we know, can't help but keep track of their troopmates and the constant tidal shifts in their relationships.

Social stress also increases with more actors. Primate research has studied "stress" produced by social instability, particularly in species like baboons whose males transfer between groups. The baboon evidence shows that when a new male joins the group, upsetting social and rank stability, all the group's males experience elevated stress hormones.[13] Other research indicates that changes in group membership, or changes in rank, as well as the presence of unfamiliar males, increase social instability, which creates higher stress levels for everyone. This happens with or without aggression in chimps, hamadryas baboons, organ-grinder monkeys, and woolly monkeys.[14]

So, when we consider what life is like for a baboon in a large group of, say, 100 animals, with interactions happening all day long—many little disputes about many little things, males coming and going, mating games, shifting alliances, food to be found, all of which require attention—it makes perfect sense that a baboon would want less of it. An overtaxed brain takes from everything else. It's a fact that the brain consumes the most energy of any organ, and the frontal cortex is hungriest, both as it develops and when it's processing information and making decisions.

It also makes sense that a heavy cognitive load and heightened stress would exact a toll on baboon performance, both collectively and individually. I had seen how uncomfortable baboons are with social ambiguity, most tellingly when the Pumphouse female hierarchy blew up and the troop's normal functions ceased until order returned (chapter 10). And I had seen baboon decision-making grossly impaired, as when Namu, a very stressed troop, invited repeated mobbings by Sisal, at great cost to their well-being. Baboons don't need to be conscious of these associations because they "experience" them through the brain's melding of cognition, emotions, and actions.

Female rank looked to be another factor in the spate of Pumphouse fissions. In translocated Pumphouse, Amua, the lowest-ranking female, ascended to the top spot by joining a group composed of local females after a near fusion of the 2 groups. And, as we saw, Pumphouse females attacked the highest-ranking family after its members joined male friend Latour at the edge of the troop.

Back before translocation, I'd already seen disruptive interactions in Pumphouse. In one particular commotion in the past, an adolescent male chose a higher-ranking family to dominate as he graduated from the female hierarchy. He'd targeted a certain subadult female and appeared to have gotten his rank above her, but this outcome was not to her liking. She waited until her adolescent brother (who was bigger than the other adolescent male) was near, then screamed for help. Her brother duly chased away the cheeky upstart. Yet he was back with another attempt once the brother drifted off. Events like this serve to illustrate how multifaceted baboon social complexity can be. This single incident involved a male testing his maturity and his rank, put in motion female resistance and family loyalty. These were males and a female engaged in layered dynamics that had consequences for the entire troop.

My impression was that the Pumphouse splinter groups actually perceived the costs of social complexity, and that, perhaps, lacking human tools for sorting things out, had found ways to minimize its toll. If so, this would be the first evidence from wild nonhuman primates that individuals may choose to leave a larger group in order to reduce their exposure to social complexity. But this can only happen under unique conditions, where food is plentiful (reduced competition) and sleeping sites are near enough to accommodate different troops (heightened predator protection), factors in play during the Pumphouse gyrations.

---

If I'm correct, baboons not only negotiate their social group but sometimes also wordlessly deliberate its size. In this way, the group's dynamic balance can and does change, in a process accelerated by the Human Age and its rapid alterations. The best test of the model would be its fit to Pumphouse baboon fissions and fusions. Unfortunately, I can't test my model on the same data I used to construct it. That would be circular reasoning. Instead, I will have to wait and watch, hoping that my ideas about the social group will not only help explain what I witnessed in Pumphouse but might also apply to an understanding of future events.

---

The thrill of science is when a blurry picture suddenly snaps into focus, when you're able not only to explain what has happened but predict what might happen. In the 1960s I was taught that primates live in social groups because it is adaptive—prima facie—and I accepted this assumption as fact. Now I am offering something quite different: a peek into the multidimensional process that guides baboons as they decide in real time whether to stay with a group or go. Unlike previous studies that used comparisons across primate species, my argument uses only one species, olive baboons. But if you accept my interpretation—that a baboon group embodies a nuanced balance of forces that entail *all* the costs and benefits claimed by others who've considered such things—then it might apply to groups of other primate species.

The dark days are gone, when I thought baboon complexity had doomed Pumphouse to chaos and possible dissolution. I benefit from the Human Age because the perspective it affords lets me see how baboons continuously construct their society, using tools they command and responding to forces, both social and ecological, they can't control.

---

What had the baboons taught me so far? Many things. The social group is a major adaptation for primates, including baboons. While the group, ultimately, is held together by the evolutionary interests of survival, relationships are the glue of daily life: friendships between females and males and between infants and males, sexual behaviors, and the play of youngsters. Males come and go. As they assimilate with the group, they create partnerships that serve as social currency in situations that require leverage. Friends are behavioral kin, which means that baboons have taken the first step *beyond* kinship by breaking the tight link between familial kinship and behavior. With unrelated animals they create "bonding capital" and thus share a human quality.[15] Baboons have more options than we ever imagined. They can choose friends and decide between aggression and social alternatives when conflict flares. They can opt to be a raider, to help create or prevent a fusion or a fission, and they can join another troop. The group also has options, visible in troop movements, range shifts, and the way small groups solve the nighttime challenge of keeping safe from predators. But these choices are never independent of the social and ecological context. A baboon group can't be understood as the sum of deterministic parts or by using reductionist theories. Social relationships, history, contingency, socioecology, and the growing human impact on the environment all matter.

---

The baboons' world was changing, as was mine. My reconstructed back gave me a new life, albeit with limitations. Now that I felt stronger, I was more self-assured and learning, to my surprise, how body and mind connect in ways I had never imagined. I was spending more time with the baboons, more enjoyably, even though most of the fieldwork was now in the hands of research assistants. The Baboon Project had become one of the longest of any primate study in the wild. I could compare many troops simultaneously or the same troop over many decades.

I am no longer a watcher of the daily baboon soap opera, but my long view lets me see "patterns" across seasons and across troops. It's a perspective that has enabled me to contrast historical changes in both landscape and animals, and to discern and decode baboon complexities others have missed. Comparison is a powerful scientific tool, used since Darwin's time. And even though I hate the Human Age, it has given me an advantage. Rapid humanized changes in the landscape have allowed me to track baboon responses in real time, liberating me from speculations using evolutionary time.

I hope by now you're convinced that baboons are much more than they seem. Perhaps you might even find them more appealing than expected, given their usual unflattering descriptions, past and present. Over five decades, as I learned what it means to be a baboon, I avoided the very question that prompted my first study: What do baboons tell us about human evolution? My framework has changed considerably since the 1970s, when I had specific reasons for thinking baboons might help us understand the earliest humans. Later, I realized I'd become preoccupied with baboons for their own sake. At this point, it seems highly unfruitful to me to use specific species as substitutes for humans, because primates—and most animals—are more intricately adapted to their conditions than any species model allows.

Given my rejection of primate species as models for early humans, you might wonder about this anthropologist's rationale for studying baboons. I nicely sidestepped this question when I decided to unravel everything I could about baboons and then focused on their conservation. Now I'm finally ready to consider the value of baboons for thinking about human evolution.

PART III

# Parallel Worlds

Chapter 14

# Why Baboons Are Not Human

## The Matter of Mind

TIME PASSED and the baboon model became a "dead issue" for me, as my Berkeley mentor, Sherwood Washburn, might have put it. The baboons had convinced me they were "almost human." The baboon social strategies on daily display clearly required acuity and skills that we'd only ascribed to humans before. As I fell further under their spell, I stopped thinking about baboons as proxies for humans or subjects of a "model." I wanted to understand baboons as well as possible, on their terms, apart from behavior they shared with humans. Accordingly, my focus shifted.

Now, fifty years on, I find myself revisiting human evolution from the baboon point of view. This time I'm struck by a question I'd never thought to ask. If baboons are almost human, why did they *not* become human? Of course, evolution doesn't work that way, and we humans flatter ourselves if we imagine our species to be the goal of evolution. But with insights gleaned and the help of findings in several fields, I can look at baboons and humans comparatively, yet separately. I began a thought experiment aiming to shed light on the divergent evolutionary paths of baboons and humans. I started with the idea of mind.

Though I feel I know baboons very well, I can't claim to have access to the inner workings of their minds. I will never know how a baboon truly perceives its world or what thoughts it might have about other baboons or the place where they live or how it all fits together. But new approaches to the study of the human mind have helped me see that "cognition" and/or "mind" very likely resides not only in our heads but in society as well. It's a concept that gives me, if not a window, then at least a peek into the baboon mind.

In the 1980s, cognitive scientists like Ed Hutchins and Don Norman, both at UC San Diego, embarked on a reevaluation of intelligence, cognition, and thought. Hutchins's novel approach was to stretch the "unit of analysis" beyond an individual's brain/mind to the social context. He called the concept distributed cognition.[1] I liked "D-cog" because it focused on what people actually do. One not-so-surprising finding was that people usually collaborate to solve problems. More exciting to me was the idea that people working together are able to "think" thoughts that wouldn't occur to a solo thinker.[2] This pointed to the importance of the sociocultural context for humans, because cultural tools like paper, pen, calculators, star maps, and computers are as much a part of human cognition as interactions with other people.[3] These are called cognitive artifacts. Humans exploit these tools ubiquitously and naturally. When we experience distributed cognition or deploy cognitive artifacts, we do so, perhaps ironically, without a thought. And yet every discussion, every instance of teamwork, and every word scribbled or typed is an example of this extended cognition, also known as "extended mind." Thus, the social group and cognitive artifacts enhance and expand an individual's cognitive process.

I applied these principles to baboons. Though baboons don't have culturally produced artifacts, they do have a large, meaningful set of behaviors and relationships—what I call nonmaterial artifacts.[4] By expanding the unit of mind to include multiple minds, the D-cog framework allows me to document and explore baboon "thinking" from a fresh perspective.[5] Since baboons lack material assets, they must traverse both the social and the ecological simultaneously, using only bodies, retained knowledge, and each other. We called this "navigational intelligence" rather than social or ecological intelligence, combining as it does elements of both.[6] Consider a female who wants to get to a specific food patch. She must also negotiate her social world to get there. That means giving wide berth to those who rank higher or perhaps displacing those who rank lower.

A deeper dive into cognition research led me to other intriguing theories that aligned with my impressions of baboon cognition. The first was proposed by Russian psychologist Lev Vygotsky, whose work in the early twentieth century went largely unnoticed until the 1970s. Vygotsky studied cognitive development in children. He found that social interactions and the social context were critical to how children learn to think and acquire abilities like walking, playing, and using language. His work culminated with a conjecture: the existence of a "mind in society."[7]

Vygotsky's framework offered a strong interpretive tool for baboon minds, with its notion that humans first solve problems through social interaction and, later, absorb the solutions as mental "schemas." I could use this approach to address something I'd long wondered about. The first time an adolescent male baboon recruits an infant as a buffer against aggression, is he, in essence, "inventing" the behavior? Did he have an "aha" moment of strategic wisdom? Using Vygotsky, I could imagine that this young male likely saw others use babies as buffers. Then, when the need arose, he decided to try it himself. Perhaps not successfully at first, as we saw in chapter 2, but he'd refine his technique with more observation and good old-fashioned practice. Over several trials in the real world, the strategy then lodged in his mind for future use.

This is the reverse of how mental schemas are assumed to work. The assumption has been that problems get solved in individual minds, followed by solutions acted on in the world. But I was intrigued by the idea that baboons, like Vygotsky's children, seem to "think" first with social interactions, then create mental schemas of strategic behaviors. Vygotsky gave this process a name: "appropriation." But because baboons lack language, the analogy is imperfect. Vygotsky recognized the special role that language plays in the development of higher cognition in humans. Humans and baboons may learn similarly, but the results will lack equivalence due to baboons' signal limitation, their inability to talk. Still, baboons appear to translate their observations into action in a way that looks to me very much like appropriation.

The second approach, called "situated action," posits that context—particularly the material environment—guides the actions of individuals in their world and with each other.[8] Vygotsky noted that humans learn to use objects by interacting with them. Situated action qualifies this by claiming that not all actions are possible because objects have properties that suggest how they can or cannot be used. For example, a pencil clearly has a job quite different from that of a can opener. These "traits" of inanimate objects, called "affordances," guide humans to behave in a certain way. Think about how you drive a car or enter your house. You can't enter a house the way you drive a car.

You might wonder how this is relevant to baboons, who do not have material culture, though they increasingly live in ours. Their world is largely composed of ecology and relationships, features that I began to see as equivalent to "objects" for baboons. Baboons may not live in a material culture, but just as we take cues from objects in our world—from our shoes, cars, guns, and microwave ovens—baboons take cues from their ecologies through situated action. Indeed, it's a matter of survival.

For example, the translocated troops followed and watched as indigenous groups found and ate acacia flowers. Later, using situated action, they set out on their own to find blooming acacia trees and feed on the flowers. Game trails, which the baboons preferentially follow, are not simply a common feature on the savanna; they have utility, guiding baboons easily and quickly through the landscape. Or consider *Opuntia* fruit. First the fruit has to be separated from the plant; its attachment point suggests how the pressure should be applied. Then the glochids, the little annoying hairs, have to be removed by rolling the fruit on the ground or on the hand. If the fruit isn't very ripe, the peel must be removed before the inside can be eaten. Sansevieria, the indigenous plant new to the translocated baboons, posed a greater challenge. First, using D-cog, they observed local baboons harvesting the plant. But preparing it for eating required clues from the plant itself, that is, situated action. A stalk must be pulled from the ground and then dexterously stripped of its outer layer. Inside lies a juicy, watery base.

Thankfully, in novel situations baboons respond creatively, resourcefully, and sometimes surprisingly.[9] I saw this time and again as the baboons adapted to their new home, and many of these new achievements seemed indicative of both D-cog and situated action. Here's a charming example: When the baboons and I came to a dirt track used by trucks and cars, the baboons would stop and "look both ways" before crossing. They had taken cues from each other and clearly sensed the risk posed by motor vehicles. From local baboons, they'd also learned how to find water in sand rivers, an indispensable skill that got them through Mukogodo's terrible droughts. Certainly, these accomplishments match many of those chimps and other apes studied in the wild and touted as exceptionally smart. Unlike the apes, however, baboons have yet to start using tools, perhaps because they've managed so well without them.

Distributed cognition, Vygotsky's mind in society, and situated action all suggest that it is wrong to think of mind as something sitting by itself inside your head. I think it's far more plausible that social and situational factors have outsized roles in how our minds develop and work. These new perspectives mean that baboon "minds" find expression in their interactions with each other and their natural surroundings. Made visible in behavior, baboon minds become more knowable.

---

As I reevaluated baboon—and human—cognition using the extended mind idea, I sensed I was closing in on the answer to a question that's haunted me for years: How do you understand a mind that is so similar yet so different?[10] This line of inquiry led me to wonder if, like humans, baboons have a "theory of mind." That is, do they have the ability to grasp that other individuals have minds that harbor wishes, desires, and intentions. The philosopher Daniel Dennett proposed a classic view of theory of mind, predicated on levels of intentionality.[11] Dennett's scheme has four levels, and here I ask for the reader's forbearance. Even I find it difficult to wrap my head around the later stages. Level one is basically: I know that you know. Level two is: I know that you know that I know. Level three is: I know that you know that I know that you know. Level four, naturally, is: I know that you know that I know that you know that I know. Dennett suggests that humans, possibly excepting me, engage in all four levels of intentionality. This means that they have a theory of mind that causes them to modify their actions in response to the perceived mental states and intentions of others. He claimed that other animals, even nonhuman primates,* do not factor in "the other's mind or intentions" before taking action. That is, they don't have a theory of mind.

Since we have yet to tap the thoughts of animals, theory of mind remains a controversial topic among animal scientists. But there's tantalizing evidence that some animals may be capable of recognizing intention in others. Language-trained, "humanized" chimpanzees, for instance, have been tested in a variety of ways. One found that some chimps will spontaneously help a researcher grasp an out-of-reach object, which means that the chimp understood the intention of the researcher. A pioneer in this line of research, David Premack has a provocative explanation.[12] He suggests that simply treating language-trained chimps as intentional beings causes them to acquire a theory of mind. I would add that this seems to occur *through* interaction. But as Premack asks: Who taught the first teacher? There is no answer. When or how our minds got theory of mind is anyone's guess.

The humanization process also elicits, or creates, a notion of "self" that we see in language-trained apes who pass the "mirror test." In this exercise, the face of the animal is marked with a dot. A mirror is then placed before the animal. If the animal touches its face after seeing its image, it has shown self-awareness and passes the test. Of course, training to mirrors is important, and monkeys, including baboons and apes in the wild,† fail this test.[13]

Recent evidence suggests that Premack is correct. It is the *process* of humanization that appears to change the brains of apes that pass the mirror test. The same seems to be the case for monkeys. In

*Dennett watched vervets briefly with Robert Seyfarth and Dorothy Cheney, the noted primatologists who also studied Chacma baboons in Botswana.

†I have recently seen reports of fish and mice passing the mirror test, but only social mice seem to have a sense of self. One wonders what to make of this.

experiments conducted by Atsushi Iriki of the RIKEN Brain Science Institute in Wako, Japan, captive Japanese macaques (*Macaca fuscata*) taught humanlike skills develop more humanlike brains. In his experiment, the researchers trained monkey subjects to use a stick tool to collect fruit placed outside their cages. Scans of the monkeys' brains pre- and post-training showed reorganization in certain parts of their brains.[14] I'm intrigued by this finding because it seems to support my idea that the baboons' social and ecological contexts mold their behavior and, by extension, their brains, suggesting a pathway toward more humanlike behaviors. In chapter 20 you'll see intriguing evidence of behavioral changes in an unnatural context, Cape Town, where urbanized baboons run rampant.

Baboons can also unlearn things, meaning they're able to replace old expectations with new ones. Today the Pumphouse baboons casually mingle with elephants. You'd never imagine that this troop, years earlier, had fled their sleeping rocks in the middle of the night after an uninvited visit by elephants (chapter 8). Gradually, Pumphouse realized elephants meant no harm. Their fears tempered, Pumphouse altered their expectation of risk from elephants and adjusted what I call their "theory of elephant behavior." I think baboons have many theories of behavior, that is, expectations based on interactions with other baboons and with other types of animals, both wild and domestic, in the ecology they all share. Baboon expectations come with the ability to anticipate rather than simply react to things they encounter.

Theory of behavior is not the same as theory of mind. But maybe baboons have a glimmer of the latter. I am convinced that baboons recognize motivation and intention in *humans*, a step beyond predicting behavior. What's not clear to me is if baboons perceive motivation or intention in their *interactions* with humans. Motivations induce us to act in certain ways, while intentions entail a plan or goal of action. In the crop raiding days at Kekopey, men were the most effective in chasing the troop away from crops, women less so, and children not at all. True, the men were larger, but they were also more aggressive than women or children. The baboons seem to sense that men meant business and, accordingly, may have "read" the men's motivations or intentions. Yet after translocation, pastoralist children were able to push the troop away when, and only when, they turned baboon deterrence into a game that was consistent and *persistent*. Had these children's determination convinced the baboons that they might as well leave?

Perhaps more telling were the baboons' varied reactions to Maasai warriors. In their traditional dress, with hair plaited and ochred and spears in hand, these men cut a fierce figure. On one occasion, I saw the group run in fright at the sight of just one warrior approaching from a distance. But I'd also seen baboons barely react as a warrior strode through the middle of the group. In the first case, I could tell that the warrior was poised to chase approaching baboons from his boma. In the other, the warrior was simply strolling from one place to another. If theory of mind depends on "knowing" that others have a mind, then I have no evidence that baboons possess this human trait. But if mind also exists in interactions, and if baboons perceive motivations and intentions when interacting with humans, then perhaps baboons do have rudimentary theory of mind.

---

Humans are clearly different, with resources that by comparison seem almost limitless. Bolstered by language and material culture, the human version of extended mind is undeniably remarkable. The interplay between human minds and all that lies outside finds us in a world of our own making, a constructed world that shapes human interactions and their outcomes. I believe baboons can think thoughts together they never could think alone. But when, in the same process, we humans harness our abundant inner and outer resources, there's a ratchet effect, with past achievements serving as springboards to greater ones.[15]

For baboons, the mental effort required to manage the intricacies of their socio-ecological lives makes them smart. For example, when two males fight about their dominance status, it's not an isolated event. There's dense context around their dispute: male alliances, friendships, agonistic buffering, competition for resources and receptive females, and more all come into play. Others watch, interested, because the outcome may affect *them*. Friendships and alliances are subject to change, giving each interaction the potential to reshuffle relationships and group dynamics. Baboons live lives that I have termed *socially complex* because their relationships and group dynamics never stop impinging.[16] I've often said that baboons need eyes in the backs of their heads to keep track of their shape-shifting world, though they take it all in stride, to a point. With all a baboon needs to keep track of, I can imagine a small chunk of a baboon's day in thought balloons: *Don't pester my friend! Will you help me? Should I help you? Will mom have my back? Hey, you can't get away with that!* This is not just noise for baboons—it's crucial information, noted and interpreted continuously. Only grooming brings social respite, but even grooming sessions end abruptly in response to a sudden movement or the noise of other social interactions. Baboons don't have vacations, though I think they might like one.

Here is a silly but pointed example of how humans and baboons operate differently. I can lead a seminar discussion about *Homo erectus* fossils without thinking for even a moment about the students' relatives, friends, lovers, socioeconomic status, where they've chosen to sit, or what they had for breakfast. If I were a baboon leading this seminar at Baboon U, there would be no lecture notes and I would strain to assess all those things about the baboons sitting before me. Moreover, I'd need to determine how *I* fit into the complex scheme. As my thinking crystallized, I realized the importance of this very human ability to *simplify* social negotiations—with the help of a material culture that includes signifiers like tools and clothing as well as language. This attribute enabled protohumans to move from a complex, baboon-style society to an endlessly faceted yet highly organized human society, the subject of the next chapter. In other words, social disentanglement helped liberate the human mind.

---

Any discussion of mind requires more than a passing consideration of that most human attribute: language. When I think about language, I think of my first child and the insight that came from watching her learn to talk. Carissa's early stabs at speech, more garble than babble, had Jonah and me convinced she'd never utter a word, let alone a full sentence. Her default method of communication was finger pointing. It was on us, in good baboon fashion, to discern what she wanted from the context. Then she finally said her first word, then another, and then another, until it was impossible to shut her up.

Up to that point, I'd been wildly impressed by language experiments with great apes—chimps, bonobos, gorillas, and orangs. But with my new maternal perspective, the limitations of ape cognition suddenly seemed stark. Whether using use plastic cutouts,[17] rudimentary sign language, or a specialized computer keyboard, apes still struggled to learn more than 200 words after thousands of hours of training. An exception was Kanzi, a remarkable bonobo (pygmy chimp) who conversed with investigator Sue Savage-Rumbaugh on a computer keyboard.[18] Kanzi was never trained, but he learned from watching his adoptive mother, who'd been a less impressive study subject. He attained a vocabulary of 3,000 words, yet never constructed a sentence of more than several words. There was no comparison between Kanzi and Carissa, whose ease of learning led inevitably to the fullness of language. Language training reveals glimmers of language ability in great ape brains, but even endless prompts and rewards fail to produce the biology of language I saw in my daughter.[19]

Language, obviously, is a convenient and efficient way to communicate.[20] Perhaps most fundamentally, baboons' inability to talk to each other is what keeps them from becoming human. Instead, baboons must act out each need and each intention, large or small. To be sure, they communicate about how they feel, but not in a way that allows the sharing of deep thoughts, regrets about the past, hopes for the future, or even the whereabouts of a great new feeding spot. It's all "show" with no "tell." Language lets humans talk about the past, mull over the present, and do something no other species has done: anticipate the future and plan for it. As Richard Dawkins said, "Humans invented the future."[21] And we are *Homo prospectus*, humans who think about the future.[22] Without doubt, a basic building block of human society is the ability to plan, and language makes it possible. The day will never come when baboons discuss tomorrow's group movements or debate the flavor profiles of cactus fruit and acacia blossoms.

Language may give humans the edge, but baboons convinced me that you don't need language to "think." Historically, philosophers, linguists, anthropologists, and various other experts have assumed that language and thought are inseparable, that words are the sole vehicle of higher cognition. This may explain why symbols, considered protolanguage, play such a central role in scenarios of human evolution. The ability to connect a thing with its symbolic stand-in requires thought. But this presumed link between language and thought means that wordless animals have unthinking minds. Yet, I see baboons thinking every time I'm with them, in their interactions with each other and with their ecology—both perfect expressions of extended mind. I see thought in their social alternatives to aggression and defense, in their clever consort tactics, in foraging and troop movement decisions, and in the calculations behind their intricate social webs.

Also, if a baboon were to forget how to harvest a corm or react to a displacement, it would only need to look around to see others doing what it might have forgotten. By contrast, chimps, with their fission/fusion society, often find themselves alone or in small family groups. Orangs are even more isolated in their semi-solitary existence. Neither species can unfailingly rely on others to help them "see" what they don't know or have forgotten. For baboons, on the other hand, the group serves as a reliable information source. While I'd expect extended mind to apply to all primates, I think it quite possible that chimps and orangs need to store more information in their heads because they lack the kind of knowledge repository provided by a baboon troop.

Of course, humans have the greatest memory aid of all: history. I thought of this when I opened an old baboon folder labeled "Inter-troop Aggression." I was surprised when the events on the page became fresh again, as if they'd happened yesterday, not several decades earlier. Somehow, my notes translated "soft" baboon history—a spate of interactions involving food and social partners—into something hard and durable that traveled across time and space and landed before my eyes. I had to wonder: Where do baboons keep their history, having no record of what they were up to last week, last year, or at any point in the past? But clearly their behaviors are based on some sort of memory of past events, not just genes. Perhaps, I thought, baboon history is embedded in their relationships and carried across generations, with transmissible traces stored in their minds. But relationships and minds can't share hard information or history. Only my notes made soft baboon history into hard historical facts, and that required language, our most powerful tool.

---

What then kept the baboons from becoming more like humans? My guess is that context made the difference. Humanizing the context changed behavioral outcomes for both language-trained chimps and Japanese macaques. These examples suggest that if you place nonhuman primates in a human setting, they begin solving problems in a novel, more humanlike way. Interaction even changes how their brains are organized—not so surprising given what we know today about brain plasticity.[23] The earliest emergence of the stuff of human culture—tools, symbols, and language—would have changed how humans interacted with each other and their environment, a "self-humanizing" process that boosted human brainpower.* Eventually there were teachers and students, experts and apprentices. Was there a cognitive Big Bang? Maybe, maybe not, but we know with certainty the ratchet effect accelerated and intensified human cognition over time.

So, the human mind produced culture and culture made humans even smarter. But baboons are pretty smart too, and as a species they've thrived, more so than their brighter ape cousins. Do baboons perhaps have their own version of culture? If so, what part might it play in evolution, along with mind, behavior, ecology, and those random curve balls no one saw coming?

*Some social scientists call this a "self-domesticating" process.

Chapter 15

# Why Baboons Are Not Human

## Culture and Evolution

**STUDENTS HAVE SOMETIMES** asked if I think baboons are on their way to "becoming human." The question made me wonder if I might be overselling baboons, just by describing what they do. After all, they *are* remarkable creatures. On second thought, the question struck me as amusingly naïve, and it began to look like an expression of a deep human bias: the assumption that humans are the pinnacle—perhaps the target—of evolution, and that all species would "want" to be *Homo sapiens*. The bias may stem, in part, from the fact that humans created capital *C culture* and therefore consider themselves special. Thankfully, in fits and starts since the 1950s, a daring subset of scientists have explored the possibility of culture in nonhuman animal populations. But before putting baboons to the culture test, let's examine what we mean not only by "culture" but also by "tradition" and "society."

The terms "tradition" and "culture" tend to be used interchangeably, referring to variations in social behavior, institutions, belief systems, and practices, across groups or populations, transmitted by social learning. Society—loosely defined as a large or small group of people sharing a culture and/or a place—is the human vehicle. But within these definitions there's debate. For example, what exactly counts as culture? To my way of thinking, as *contexts* change, so too does behavior, no matter the group. For our earliest ancestors, contextual change included the emergence of material culture: first tools, then symbols, and later, language. When humans branched out across the globe, creating cultures as they spread, they provided the means to differentiate between groups. Those distinct cultures in turn shaped the people who lived in them. Thus, divergent groups of our species—for example today Americans and Kenyans—fashioned distinctive ways of doing things. Events in both real and evolutionary time offer modifying opportunities and constraints on the future. As humans manipulated their surroundings to expand their unique niche, both process and outcomes gained velocity and entrenchment.

The name Bruno Latour will be familiar from earlier chapters. Early in our careers, in 1978, I invited Bruno to my first Wenner-Gren International Symposium, titled "Baboon Field Research: Models and Muddles."* He was working at the Salk Institute, not far from the University of California, San Diego campus (UCSD), conducting a study of scientific practices at the Roger Guillemin lab.[1] That made him the ideal participant in any discussion about how we do science and why we find some answers more satisfying

*I recount the surprising outcome of this symposium in chapter 18. The Wenner-Gren Foundation for Anthropological Research was founded and endowed by a Swedish entrepreneur in 1941. Wenner-Gren symposia are organized periodically by anthropologists around emerging issues in the field.

than others. Thereafter, Bruno and I created and for several years co-taught a university class, Material Culture. I found him to be disarmingly affable as well as a brilliant, original thinker. We remained close friends until his death in 2022. His gift was to see beyond the obvious, then explore what he found there.

At first I had trouble understanding him, aside from the French accent. But once I caught on to his way of thinking, I realized my own perspective was shifting. When he asked, "What is this thing we call 'society'?" I was shocked by the question's simple profundity. We assume "society" has always existed and take it for granted (see chapter 13). But somehow societies get built—so, by whom, and how? Inspired by pioneering work of sociologist Erving Goffman, Bruno argued that society isn't something "out there," a fossilized structure into which humans are born.[2] At Salk, Bruno observed research in action, taking special note of the tools and collaborations that enabled science to move forward. Extrapolating from those interactions, Bruno postulated that human society, broadly, results from a malleable and negotiated "process," and as such undergoes continuous re-creation.

In our work together, Bruno and I went a step further. Bruno visited in Kenya, wanting a firsthand look at baboon society. He saw what I'd been observing for years, countless daily negotiations among baboons about everything from troop movements to baby sharing to rank challenges. He was, as they say, blown away, not only by the baboons' social give and take, but by the surprising resonance with his lab observations. Both societies were products of negotiation and process. But there was a problem. If both baboons and humans continuously create their societies through negotiation, then why were the outcomes so unalike?

Bruno and I theorized that this gulf arose because baboons and humans created their societies using different resources. Baboons have only their bodies and social interactions. A baboon's body carries vital information about its age, sex, and condition that other baboons readily perceive and interpret. But it doesn't stop there. Baboons assess each other's social lives as well—rank, kin network, friends, and social interactions. I see them take note of what another animal chooses to eat and where it sits in the group. Many aspects of baboon life are built on assessments and decisions that I infer from watching their interactions. And what I've seen tells me that a baboon's existence entails moment-to-moment effort, deliberation, and maintenance. Thus, Bruno and I arrived at the term "complex" to describe their *society*, because so many entangled variables simultaneously bear on a baboon's decision about what do next.[3]

Let me illustrate, literally, a key difference between baboons and humans. Imagine if baboons wore hats. A hat signaling a baboon's likely behavior would be, in essence, social shorthand, a material clue and cue that transmitted information and helped others predict what the hat-wearing individual might do. One might be a baseball player or a safari guide, another a nurse or even Santa Claus. The hat would be the first step—material culture—in changing a baboon society into a human society. Adding something as simple as hats to the baboon repertoire would simplify their social negotiations (plate 20), as they do in human interactions. Of course, this was my trick.* Baboons don't have full-blown material culture. No nonhuman primate does.

This is a new way of understanding how human inventions help us cope with the messiness of real lives. All sorts of material markers accumulated as humans evolved—not just hats, of course, but tools, dwellings, cars, money, and everything else. These *things* helped human culture, in all its forms, transcend the limitations

*I used a replica of a baboon head from a 1990 photo exhibit at the San Diego Museum of Man. Unfortunately, it was of a chacma baboon (the southern variety) and not an olive baboon male. Still, the hats covered up the greatest difference between them: their foreheads.

of a society whose members have only behavior to explain themselves and whose acted-out negotiations mire them in social complexity.

Humans have forged comparatively strong and resilient social units like clubs, schools, churches, towns, states, nations, and multinational corporations. While social negotiations among individuals continue at the micro level, the macro units can be stabilized. We've even managed to transform objects into social actors, making hybrids of real people and material surrogates.[4] Two of Bruno's favorite examples—quaint as they now may seem—were speed bumps and automatic doors. Speed bumps (the French call them "sleeping policemen") take the place of a person slowing traffic; automatic doors replace a person who opens and closes doors. Substituting objects for people further simplifies negotiations. After all, there's no arguing with a speed bump or an automatic door. In ways like this, humans codified their behavior, economized and stabilized their social negotiations, and thus were able to scale up their society.* By contrast, baboon troops can splinter when they grow so large that social (and foraging) challenges reach a breaking point (but see chapter 13). And while female matrilines provide a modicum of continuity in baboon society, male baboons barely maintain fragile alliances, and then move on.

Our material culture and language help us segregate our lives into myriad strands, while baboons lead lives where everything, everywhere, is all at once. That is, our attention is divisible, not dispersed as it is with baboons; we can focus on one thing or another with a minimum of distraction. Bruno and I decided to use the term "complicated" to describe human lives and society. Think of the amazing calculations a computer makes using the simplest of units, or the wildly varied yet basic crystalline combinations that form

*Which isn't to say that human society is free of conflict and instability, despite its stabilizing institutions.

snowflakes. Both are complicated but composed of discrete elements, like our lives.[5] Humans live our complicated lives so naturally that we're barely aware of how complicated they are. When our human challenges impinge from all sides, as they do every day for baboons, we are easily overwhelmed and sometimes impaired.

---

Nonhuman primate groups are the most obvious place to look for culture in animals. The earliest evidence of nonhuman primate culture came from research in the 1950s and 1960s, on wild but provisioned Japanese macaques.* Japanese scientists concluded that these monkeys had "proto-culture"—a set of inventions passed between individuals through social learning.[6] The monkeys were observed washing sweet potatoes in the sea as a way to salt them, and cleaning sandy wheat by tossing handfuls onto water—practices that persisted over time. This was a thrilling discovery, the first documented evidence of culture originating in a nonhuman animal. But the bubble burst when the scientists recently learned that the monkeys' handler was rewarding the behaviors. That meant a human had created this monkey proto-culture.

Half a century later, a very different sort of study was undertaken. The result was a compendium, completed in 2000, of behaviors observed in wild chimpanzee populations dispersed across Africa. The data yielded convincing evidence that chimps met the scientific criteria—at that time—for culture.[7] The report noted differences among chimp populations in behaviors like tool use and grooming. There followed a comprehensive review of orangutans in the wild. It concluded that they too had culture, seen in variations between groups in styles of harvesting and preparing food. These distinct techniques could not be explained by evolution

*The same Iriki monkeys who, in the previous chapter, were taught to use sticks to reach fruit outside their cages.

because the slow rate of adaptive change couldn't explain the variations between groups.[8] A few other primate species have joined the culture club, including tool-using capuchin monkeys in Central and South America.[9]

But do baboons have culture? They have traditions, defined at the time of my first study as behaviors passed between generations by social learning. I thought I'd witnessed the birth of a tradition in 1973, when those several Pumphouse males joined Rad in the hunting of young Thomson's gazelles. The earth shook in the world of anthropology when I published my observations. Back then, traditions signified "culture" and baboons with culture were unheard of. Then Rad moved on to a new troop and the hunting team disbanded. The episode proved an aberration, not a budding tradition.[10]

Several years later, still on Kekopey, the advent of crop raiding offered another test case. I had the perfect subjects: naive baboons tempted by something new in their world—field crops. Some Pumphouse baboons seized the opportunity, becoming raiders, while others did not. Would the raiders continue raiding? Unfortunately, they did, for nearly five years. Clarity came when I compared hunting with crop raiding. There were key differences. Hunting on Kekopey was rare, involved only a few animals, and happened away from the troop. That meant most of Pumphouse only saw dried blood on the returning males. Crop raiding, by contrast, was a daily affair, visible at times even to those who didn't join in. Visibility and frequency are the likely reasons why crop raiding persisted. Crop raiding counted as a tradition because it was a new practice, embraced and passed on selectively by certain individuals within Pumphouse, who then peeled away to form their own specialized raider troop. Much as different groups of chimps and orangs display distinctive behaviors (i.e., have their own traditions), various baboon groups had adopted differing practices. In the early 1980s, that meant they all had culture.

As we planned the translocation in 1984, I counted its possible benefits, hoping they would outnumber the risks. One such benefit was the chance to see new traditions forming in a novel ecological context. I knew well the habits of the three Kekopey groups and expected they'd adjust in their own ways to a fresh environment. I wasn't disappointed. I saw the reconfiguration of old behaviors as well as the emergence of new and lasting behaviors. Diet, considered the result of slow evolutionary adaptations, may seem an unlikely place to look for rapid changes in traditional behavior. But in their new home, the baboons were starting over. At least half the nourishing foods in dry season Mukogodo were plants they'd never seen on Kekopey. Social learning helped as the translocated troops took food-finding cues from local baboon groups and the indigenous males who joined them. But on their own they soon added new plants, starting with foods that most closely resembled those at Kekopey. To my surprise, even where they chose to sleep seemed determined by tradition. I saw their displeasure with the smooth-faced kopjes on Chololo, so unlike the rocky fault scarp sleeping sites they'd left behind at Kekopey.

There were changes in baboon predation as well. I had been forewarned by local Maasai that indigenous male baboons were known to take small stock, mostly young goats. Thankfully, the newcomers didn't raid. The practice only developed after indigenous males took the place of Kekopey males in the translocated troops. Adolescent males—descendants of translocated baboons—watched and learned from the locals. Not all baboon groups preyed on livestock, but because some did, I once again faced conflict with humans. Initially, meat-seeking male baboons targeted young livestock, about the same size as the tommies on Kekopey. Later, a few ambitious males modified their technique in order to capture adult female goats and sheep.* And at the end of the dry season, I would

*I have been advised not to describe this new technique.

sometimes see a baboon male perched at the top of a rocky sleeping site, his gaze focused on a livestock herd grazing near a distant boma. But these males always hunted as individuals, not as a team. At Mukogodo, I documented over a longer period of time what was hidden to me at Kekopey: that predatory behavior undergoes gradual alterations, evidenced by baboons adding adult livestock to their repertoire. But I never saw a replication of Rad's tag-team hunting style. I concluded that the baboons brought social traditions to their new home, but also tailored their behavior and created new ways of doing things—new traditions—in response to a different ecology.[11]

Today, there's a new set of criteria used to determine animal culture. In the current model, social traditions must be integrated and sustained in a group's way of life, and they must accumulate modifications over time. This would be the animal equivalent of the "ratchet effect" that turbocharges human culture.[12] Though I would expect disagreement in some quarters, I think a convincing case can be made for baboons as animals with culture. If baboons create society, it follows that each baboon group possesses its own integrated set of transmissible traditions in a social structure that gets modified over time. Adjacent baboon groups have different social organizations, despite being built from the same evolutionary elements. For example, each group has a female and male dominance hierarchy, and these hierarchies preexist but are modified over time. Even the stable female hierarchy has families that exit or die out, as well as rare periods of reorganization, as we saw in chapter 10. Kinship groups, that is, matrilines, are the bedrock of baboon social organization, yet they aren't absolutely fixed. Families vary in size and composition at different times, and each configuration will infringe in a variety of ways on the web of social interactions.

While the basic constituents of baboon social groups have strong links to evolutionary principles, baboon ways are flexible, integrated, and transmitted by individuals across generations. Most crucially, a baboon society and the relationships within it get modified over time through social interaction. You might say it's like a single house undergoing regular renovations. In this way, perhaps baboons do have "culture," because baboon groups have distinctive behaviors, and not just socially. (See also "peaceful" Mara baboons.[13]) Some, for example, prey on guinea fowl, others don't, and plant preferences often vary as well. An incoming male needs to learn the culture of the group he's joined. He might also be altering the culture of that group.[14]

Determining which species have culture is tricky, given the mutability of the concept. But I have to wonder if baboons need more culture than they already have. As I suggested in the previous chapter, what a baboon needs to know about baboon conduct is available all the time in the form of its cohesive social group. It simply needs to look around. If culture encodes and relays information about behavior and survival, then the group itself serves as an instruction manual. And while I'm happy to make a case for baboon culture, I can't claim with certainty that my favorite animal actually ticks all the culture boxes. Which is to say, my stance is complicated and slightly conflicted. I recognize that baboons are innovative, adaptable, and good problem solvers. However, if you buy my argument, that baboon social structure is an example of culture, then we need to have a closer look at the origins of human "culture," a point I'll expand on shortly.

Of course, there's no denying the vast gulf between human and baboon cultures. Baboons have only "soft" tools—their bodies and relationships—that limit the scale and durability of the societies they build; the ratchet effect applies, but barely. If we wish to detect shifts in an animal society, then we need to pay close attention to changes in their worlds (i.e., their contexts). Meanwhile, humans have, use, and relentlessly upgrade their hard tools. They keep records, invent new things that get shared, nurture their institutions, and don't seem to know how to stop ratcheting. The results set them apart from all other life-forms. Just look around.

---

There is another answer to the question of why baboons aren't human—evolution. My thoughts on evolution include a personal conviction, that evolution is *not* deterministic. Biology plays a part, but it's not destiny, per the hoary refrain still invoked by assorted behaviorists and biologists.[15] Nor is everything possible. Humans will not sprout wings and fly, and baboons will never be Olympic swimmers. Even Darwin recognized the constraints of anatomy and physiology.[16] Counter to Darwin, recent adaptationist interpretations of evolution assume that natural selection operates on individual traits such as sharp teeth, bottoms with red sexual swellings, even mind.[17] They ignore the whole animal and its hugely important context.

Interestingly, a smart animal can circumvent certain limitations and thereby defy evolutionary expectations. Take male baboons. Evolution built the male baboon body for aggression, but this anatomy did not ordain a violent destiny. Males learn social strategies designed to avert physical conflict and they deploy those strategies when possible. That means smart animals like baboons have options. They can become raiders or not, or they can hunker down in a familiar home range or decide to find a new one. *The idea of "options" dramatically reshapes evolutionary interpretations*. Creative individuals can alter a trajectory, and not always for the better.

Consider the evolutionary implications of Namu's mistake. Seeking *stricta* fruit, they had invaded the home range of another troop, suffered years of mobbing and stress, and lost many infants and adults. It took years for them to recover and thrive once again. Namu's story means evolution does not require perfect adaptation. The strong implication is that evolutionary processes may be more lenient and tolerant than we'd assumed. Adaptations only have to be "good enough," the organism "fit enough," to get genes passed to the next generation. This relegates Herbert Spencer's "survival of the fittest" to the category of a catchy but inaccurate slogan. It also makes the history of each baboon group, or any species, the result of good and bad luck, not an orderly, inevitable march toward "improvement" or "progress" or "us." In the evolutionary history of a species, baboon or human, both chance and necessity play a part.[18]

What I'm proposing about baboons suggests another revision to evolutionary scenarios. Before Darwin, philosophers (scientists of the past) argued about the "natural" or original state of humans. But a recent and welcome sea change has occurred in the way science views humans. To wit, each species in the primate order is unique, yet similar, because our evolutionary outcomes include both a shared history and distinctive adaptations.

Furthermore, humans possess both an ancient self-interest and a newer adaptation for cooperation, called a "dual nature."[19] Jonah's well-researched perspective suggests that human singularity began when the constraints of ecology and kinship were broken.[20] An example of this "ecological emancipation" would be when humans devised ways to carry water and thus could range more widely. Working with people other than relatives on cooperative endeavors—planned, with shared goals—broke the constraints of kinship. But cooperation alone wasn't enough to create human culture. Also needed was the ability to contemplate the future and plan for it, using tools, language, and "new" social emotions—ambition, shame, jealousy, and happiness, among others. When humans domesticated plants and animals, they gained an advantage over hunter-gatherers, and what happened after domestication is another story.

I'm reluctant to speculate about human evolution,* but I agree with Jonah that human nature includes the same self-interest

*This is why I continue to study baboons.

found in all species, as well as a unique human aptitude for cooperation. Looking at humans through a baboon lens also highlights why our distinctive traits, tools, culture, language, large brain, and *cooperation* got us to where we are, far from the savanna. Baboons, as I have portrayed them, also have a dual nature, self-interest on one hand and, on the other, the ability to collaborate with unrelated baboons. But I've yet to see them draw up plans for a cactus fruit delivery service.

Today I have come full circle, but with time and experience my perspective has changed. I think about human evolution, and humans, taking into account what I know about baboons. Baboons are not merely a simplistic version of humans, nor, I think, should they be viewed as referential models for human evolution (see chapter 18). Baboons showed me the limitations of what is possible, which gave me a way to see the prodigious added value of human traits. I can now reimagine human accomplishments—tools, symbols, language, and culture, at once both familiar and extraordinary—through my baboon lens.

Years ago, I met my hero, Gary Larson, at a social gathering. I love his cartoon called "Vending Machines of the Serengeti," in which a lion selects from a vending machine containing giraffes, zebras, and gazelles. In another favorite, a female chimp finds a blonde hair in a night nest shared with a male chimp. Brow furrowed, she asks her mate if he's been doing research with that Goodall tramp again. At first, I watched the bookish-looking Larson from the sidelines, inveterate observer that I am. Then I came up with a conversational gambit. I would congratulate him on his understanding of new scientific findings that restored abilities to animals, including emotions and mind, that behaviorism and classical ethology had ignored. I worked up my nerve, approached, and said my piece. He blinked and said he just made animals act like humans and humans act like animals—that was the humor.

I wish my approach to understanding baboons could have been that simple. Baboons helped me focus on *context* and *process*, not just results. This is why I believe scientists need to think far more deeply about the role of human innovation in the building of human society. The extended mind approach speculates that discoveries can come from social interactions, later to be absorbed into individual minds and embedded in the culture. The implication is that early human history might not be about one bright person who looked at a rock and, for the first time, saw a tool. Eureka!—tools began. I think it likelier that tools came about as a group effort, so perhaps the iconic prologue in *2001: A Space Odyssey* is not so fantastical. Our past scenarios of human evolution lie within the frame of human exceptionalism, with humans, of course, inventing everything. Certainly, given what I've learned about baboons, this doesn't make sense. Before having tools, early humans would have been at least as clever and socially sophisticated as baboons. Then factor in the amplifier: "mind in society." Now consider that evolution may be more tolerant of imperfection than we assume. This opens up a vastly enlarged space for variegated adaptations, and perhaps makes imperfection the adaptive spark behind humankind. Be it thus or not, what I know about baboons suggests that we need to rethink the starting point for humans becoming human.

The baboons reconstructed my brain and showed me how much can be accomplished through interaction. They are impressively sophisticated nonhumans—indeed, almost human. When we elevate human uniqueness, we underestimate the abilities of other creatures.[21] Yet each time we grant a human quality to a nonhuman animal, particularly a nonhuman primate, we modify another criterion for humanness. During my lifetime, the Rubicon to cross for humanness was first "traditions," then hunting, then tool use and tool making, and then culture. Now the spotlight in evolutionary discussions is on language, imitation, cooperation, and consciousness. I know the benchmark will keep shifting as we gain understanding about other animals and learn to respect animal abilities that earlier scientists dismissed.

Over the decades I have faced no end of perplexities, and the journey has not been easy. I strayed from questions about human evolution, yet now find myself pondering the ways in which baboon society can inform our thinking about evolution, particularly human evolution. Reflecting on the similarities and differences between people and baboons, as I have come to know them, can take you through time and lead to thinking afresh about matters that may have seemed settled.

## Chapter 16

# Vindication

AUGUST 2018: I was chilled to the bone. No, I wasn't on a predawn baboon stakeout. I was in a Nairobi auditorium with 800 other primate scientists at the biannual International Primatological Society Congress. I had left my warm jacket behind, thinking I wouldn't need it. So here I was, shivering in a thin sweater and shawl and wondering if the reception for my plenary address would feel equally chilly.

I had selected a topic that was sure to raise eyebrows and perhaps a few hackles: "Why Natural History is Important to Science: A Baboon Case Study." In fact, I'd almost declined the invitation, then realized that my age conferred me "silverback" status; I could talk about anything I wanted. My topic was offbeat but important, and my covert agenda was to reach young scientists in the audience. They had been socialized into the type of science I planned to criticize. I would draw on my career and baboon studies to reveal the shortcomings in primate field studies using a "hard science" approach—quantitative research focused on only answering narrow questions and testing hypotheses. I would argue that these reductionist methods completely missed the complexity—the intricate social and ecological web that, once detected, unlocks a deeper understanding of primate life in the wild. Young scientists, immersed in the quantitative paradigm, would be dimly aware, at best, of the observation-based methods used in earlier primate field studies. But they needed to understand the history and the consequences of forgetting. I had less hope of persuading my peers. These were the colleagues who chanted, "The plural of anecdotes is not data."[1] Like me, they were trained to reject natural history.

Which makes my defense of natural history sublimely ironic. My first field study was a reaction *against* those early "natural history" studies of baboons that produced the baboon model and generated such controversy. Washburn felt certain that the new quantitative methods I would apply in the field would support the baboon model he'd helped create—based on natural history observations. Why all the bother about baboons? Washburn and others were convinced that baboon behavior held clues about early human evolution. But did it? I wanted my observations in the field to provide evidence, one way or the other. My original plan to study patas monkeys—savanna-living primates with their own social organization—was in fact meant to investigate *their* possible link to human evolution.

Of course, neither Washburn nor I had the faintest inkling at the time that baboons would become the center of my life in a study lasting five decades. Or that I'd wind up standing before my peers in a chilly auditorium in Nairobi, arguing for the value of natural history after a lifetime of data-driven quantitative research. The baboons had taught me that primate science wasn't an either/or proposition. Any attempt to truly understand a creature other than ourselves requires both natural history *and* quantitative science.

Let's first revisit the "baboon model." A model of baboon behavior emerged from three observational studies of different savanna baboon species: Washburn's of yellow baboons in Amboseli, Kenya; his student DeVore's of olive baboons in Nairobi National Park; and Hall's of chacma baboons in southern Africa—all in the 1950s. Remarkable behavioral similarities were noted, no matter the species or location. Males competed for dominance. Dominant males monopolized fertile females. The male hierarchy structured the group. Males defended the troop against dangers, influenced troop movements, and policed internal disputes.

But later research raised questions about the model. Thelma Rowell studied baboons in Uganda in the mid-1960s. She saw males retreat to the safety of trees at the first hint of danger. Her work dented the baboon model. Later, several researchers never saw the model's key formation—male baboons surrounding females and young in troop movements. Hints of a female dominance hierarchy, ignored in the model, emerged from a study of rhesus monkeys in the Caribbean in the late 1960s. And yet the baboon model remained the default framework for interpretations of early human evolution, a bit bloodied, perhaps, but unbowed. Remember, these were early days of primate studies in the wild.

The scientific crux was this: Was observer bias behind the conflicting descriptions of baboon behavior or was the model wrong, not just for baboons but for early humans? With just natural history observations and its taint of subjectivity, it was impossible to tell. But it's clear to us now, what the problem was with those early studies. Washburn, DeVore, and Hall had only identified and tracked individual large males. Assumptions made about females were entirely based on their interactions with male baboons. From this skewed perspective, those early misconceptions made sense: Females lacked a dominance hierarchy, estrous cycles determined female status, and infant care was a female's primary role. While the baboon model still held sway, those niggling questions weren't going away. Which is precisely why I wanted to be among the first to use a new, more objective approach. The "hard science" quantitative methods I would apply in the field would correct for observer bias. I'd track *all* the troop's large animals, females as well as timid males who were previously discounted. Was Joan Didion correct? "We all distort what we see. We all have to struggle to see what's really going on."[2] My early data-driven approach was designed to minimize, if not erase, interpretive distortions and to reduce the "struggle."

I had another reason to argue for natural history on this day. Though I began with and continue using quantitative methods—the bean-counting aspect of science—I had a hunch early on that "facts" alone would never plumb the rich complexity of baboon ways. And I suppose this is why I began making qualitative notes on baboon relationships, behaviors, interactions, and events. I was, in fact, building a comparative, systematic, and utterly unintended natural history—the very thing I'd set out to repudiate. I don't know why it took me so long to realize this, except that I was that much in the grip of my scientific enculturation. But lured by the baboons, I finally got there. It was natural history that allowed me to unveil baboon socio-ecological complexity. And I was going to argue that natural history was, indeed, science (see chapter 18).*

---

The first day of the conference was almost comically rife with technical glitches. (We had expected better of the host UN agency.) Fortunately, my talk was scheduled on day two; I figured by then the problems would be resolved. But my slides were still missing 10 minutes before my talk. Once behind the podium, I struggled with the computer's timing device. So I improvised, rushing through sections and glossing over examples, and finished before my allotted time was up. That meant plenty of time for questions. Instead, I faced silence. I assumed my position was

*In many places I use capital *S* because there are two sciences: Science is reified ideas while science is research. See also chapter 18.

too radical, too unsuited to the zeitgeist. Or maybe people just needed time for the ideas to sink in. After all, it took a lifetime of baboon watching to convert me. In any case, my talk had seemingly landed with a thud. I left the stage and went outside for tea, weaving my way through clusters of people and their unintelligible mumblings. But as I stood sipping tea, people began to approach, first a few, then a steady stream. Several senior scientists murmured "right on," "good for you," "yes, it's obvious." But the junior scientists' grateful reactions were most rewarding; these included expressions of thanks before their own talks. I had seemingly given them needed permission to accept natural history as science. Many younger scientists were studying elusive animals, not as readily trackable as baboons. They'd found themselves doing natural history by default, making notes and writing descriptions when "hard" data was literally hard to collect. They had been nervous. Their colleagues and funding agencies were still insisting that grant requests and publications be "tests of hypotheses." I hoped I'd given them cover—the maverick silverback, advocating for natural history as an indispensable adjunct of science in the wild.

---

Why did it take me so long to embrace natural history? I'd like to explain, because it says something about the nature of entrenched ideas and how hard it is to change minds. What's odd is that I was collecting "baboon stories," anecdotes, from the start, simply because I was fascinated. But I was also blinded by the power of what *Science* was supposed to be—something unfailingly objective, quantifiable, and subject to replication. Recall, the mantra at the time was, "The plural of anecdotes is not data." To push back on this axiom was to risk professional suicide. And so I gladly adopted the rigorous, data-driven methods meant to resolve the baboon model controversy.

But from the first, there was that hunch, along with my unexpected fascination with the baboons. I began noting social context on my data sheets, and I created a daily troop diary, an informal record of intriguing behaviors that didn't always qualify for the data collection check-sheets. At day's end, I'd type up these "qualitative" notes, recording not only my sampled animals' activities, but a wide range of traits, quirks, antics, feuds, friendships, comedies, and dramas that caught my eye. Data from my "follows" detected signs of a female dominance hierarchy and a surprising instability in the male hierarchy. But, in retrospect, the numbers didn't tell the whole story. Only when I combined both information sets—description and data—could I begin to evaluate the baboon model, because "facts" divorced from context lack meaning.

I was excited by what I'd learned about baboons in my first study. Washburn was noncommittal. I might have upended the baboon model with those early findings, but it was not to be after Berkeley's massive mainframe computer crashed and physically destroyed punch cards containing 16 months of quantitative data. I had to change the subject of my PhD thesis, from a test of the baboon model to a study of baboon predation. Recall that I'd witnessed several Pumphouse baboons team up to hunt baby Thomson's gazelles at Kekopey, an unprecedented behavior and a topic of great anthropological interest. Here my written notes proved indispensable, with their record of individual hunting events, the animals involved, and other detailed descriptions. The total number of captures over time suggested that these baboons were the most predatory primates yet observed—more predatory than even Goodall's famous Gombe chimpanzees. Indeed, I was convinced I'd seen the birth of a baboon hunting tradition, a discovery important enough to be published in *Science*, the preeminent American science journal. After all, coordinated hunting played a crucial part in human evolution. The idea that baboons had invented, benefited from, and might perpetuate this very activity had profound implications.

And as it turned out, my robust descriptive record served as data and rescued my future as a primate scientist.

Next came crop raiding. More than a few colleagues urged me to walk away from Pumphouse, now considered "unnatural" baboons by virtue of their predilection for farm crops, but I kept going (chapter 4). Pumphouse became a troop divided: Some baboons routinely raided, while others chose not to. I added innovative datasheets that tracked behavioral changes in this new regime, and I began to monitor the grassland ecology, allowing me to quantify baboon foraging options. I examined raiding's impact on female reproduction and kept track of how baboons used their home range. I tested raiding deterrents, including taste aversion experiments devised by a colleague.[3] The data I gleaned allowed me to compare the advantages and disadvantages of raiding and to assess its impacts on baboon life. So far so good, scientifically at least.

But my qualitative notes, those despised baboon stories, turned out to be equally important. They described the social dynamics involved in the troop's fissure and how some individuals became raiders and others did not. My amalgam of facts and anecdotes resulted in a nuanced explanation of crop raiding—its development in a group of naive animals, and its costs and benefits. The two sets of data, narrative plus numbers, were more powerful than either alone.

Something else was happening. Paying attention to everything, the whole baboon panoply, was opening my eyes in other ways. I had begun to sense the significance of an animal's history and the parts chance and contingencies play in baboon behavior. Only later would I see that my shifting perspective fell under the rubric of natural history.

---

The translocation was the next ordeal, but by then I'd stopped caring if what I was doing was technically Science. As I searched for a suitable new home, I drew on data as well as my sense about baboons after 12 years of watching them. Data and intuition both helped as I planned their capture, transport, and release, hoping to minimize risk and maximize the odds for success.

The translocation and its aftermath were major team efforts. Kenyan research assistants documented dietary shifts and establishment of home ranges after the move. I compared the translocated animals' reproductive success and survival rates to those of an indigenous troop we'd added to the study. However, I turned to my qualitative notes to provide context when the numbers failed to yield understanding. The elephant incident happened shortly after translocation, and I described it in detail (chapter 8). Without those notes, the Pumphouse range shift might have looked like a random impulse or possibly food-related, not as a search for safety. That got me thinking. Rare events like this are ignored in quantitative studies. Yet elephants had shaped the future of Pumphouse as surely as an iced-over Bering Strait led to the peopling of the New World. My interpretation of the range shift emerged from blending multiple perspectives and both quantitative and qualitative data. The team's monthly reports added depth, as did information gleaned from regular project meetings where we discussed ecology, troop behavior and unusual events. Taking all into account, including context, chance events, and contingency resulted in a complete story and a new angle on evolution.

Fifteen years went by. I began to notice changes in the environment. An unidentified cactus appeared on the savanna and began to shape baboon foraging. Though I didn't know it at the time, an "invasion" had commenced. I knew I needed to document this shift in the baboons' ecological context. I would want to know how this cactus might perturb the lives not only of baboons, but of people, livestock, and elephants. I anticipated reverberations throughout the ecosystem. Routine data collection was carried out by the field team. My notes detailed how individual baboons and troops responded to the cactus, and factored in as well impacts of seasons, people, other troops, livestock, and more. The team included a series of foreign interns, three at a time, and up to ten Kenyan researchers. We covered not only Pumphouse but its several offshoot groups and, later, Namu. The team had a close-up view of daily events, but I had the long view by virtue of my time in the field across years, seasons, and troops. I could see patterns. Knowing how past events played out gave me an interpretive framework for the present. And what I saw was a "perfect storm." Torrential El Niño rainfall in 1998, plus new human settlements, plus changes in livestock grazing—these were the change agents. The ecological constants were baboons and elephants. All combined to create cactus-friendly conditions and a distribution system. My notes helped turn plot points into a narrative (chapter 11). That's when I finally understood that my notes were systematic and comparative baboon natural history—and a boon to my science.

---

I had in fact made a rediscovery. Natural history has a long and illustrious genealogy. Starting with Aristotle and continuing through the Enlightenment, the hallmark of "natural philosophy" was careful observation and description, *not* quantification. In the sixteenth–eighteenth centuries, classification and taxonomy, based on natural history, systematically organized species and became a means of universal scientific communication.[4] Yet it was more than classification. Natural history laid the foundation for all the natural sciences. Pioneers included Charles Lyell, whose natural history begat modern geology; Alexander von Humboldt, who created the first scientific biogeography; and Alfred Russel Wallace and Charles Darwin, whose natural history produced no less than the theory of evolution by natural selection.[5] Natural history gave rise to the "scientific method," immortalized by its first step: careful and patient observation.[6] Grounded in description, emergent science advanced to predictions about studied phenomena across the span of time. Natural history spawned great theories.

So we need to ask how natural history, the origin story of so many natural sciences, became scientific heresy. The transition began in the second half of the twentieth century. In 1959, Karl Popper, a leading philosopher of natural and social sciences, published *The Logic of Scientific Discovery*.[7] In this pivotal work, he argued that a theory or hypothesis could only be refuted, not proved. Popper's scientific methodology, based on "falsifiability" and experimentation, spelled the demotion of observation-based natural history. Another blow came from John Platt, a University of Chicago biophysicist. In a 1964 paper published in *Science*, he proposed an approach called "strong inference."[8] I read the article when it appeared, but only later did I get its import. Platt argued for the superiority of certain fields of science over others and specifically urged a physical science approach to biology. That meant generating testable hypotheses, favoring simple models, and focusing on "averages," meanwhile ignoring individual variation. Biology, according to Platt, should stress experimentation and quantitative studies, and so it did, for the next fifty years. Primate studies shifted to a hypothesis-driven model a bit later, and for a specific reason.

Popper and Platt's ideas set the stage for a research revolution. Sociobiology, a perfect fit with Platt's template, wrote a new script for primate studies. Earlier I discussed how sociobiology solved the conundrum of variation between species by incorporating genetics into a new evolutionary synthesis (chapter 4). Evolutionary fitness in terms of "kin selection" could be "inclusive," meaning an individual's close relatives were counted along with its offspring; all those related genes factored into the next generation. Genetics became a feature in interpretations of behavior as well. Sociobiology's evolutionary blueprint enabled predictions about behavior that could be framed as testable hypotheses. Underlying this approach were the assumptions that individuals always compete for limited resources and "fitness" can be measured by reproductive success, defined as the number of surviving "genes."

It was an exciting time, and I was not immune to Platt and sociobiology's "physics envy." In the late 1970s and afterward, primate scientists embraced this new framework. At last, they would be doing *real* Science, testing hypotheses and producing compelling, quantifiable results. Granting agencies agreed. Natural history stopped being Science. In fact, it became the opposite—anti-science, based on unreliable stories.

So for a long while, I was an accidental practitioner of natural history, even as I assiduously studied baboons using "hard" science methods. You could say natural history was my dirty little secret. When I finally owned up to it, my sense of relief was profound. I had been setting aside my insights that came from untestable observations. Natural history suddenly legitimized them. My stories weren't mere "anecdotes."* They were firmly in the vein of Darwin's systematic and comparative natural history.[9] His natural history used close observation to detect patterns that shaped his theory of evolution. I did not mind being in Darwin's company.

I searched for other natural history icons. The first ethologists—Lorenz, Tinbergen, and von Frisch—who began modern animal behavior studies, called themselves "Curious Naturalists." For them, the evolutionary function of a behavior could only be understood in its natural context. In reaction to the post-Darwin human bias that crept into the study of animals, they had strict rules about how to collect data. Even so, they saw natural history as essential to any meaningful evolutionary interpretation of behavior.

Charles Elton, considered the father of ecology, was another natural history advocate. In the 1930s, he wrote, "Ecology is a new name for a very old subject. It simply means scientific natural

*Technically, a short or amusing story about an incident or a person; an account regarded as unreliable.

history."[10] But a review of more recent ecological work shows that ecology has shifted to a new data-heavy approach. Only a few dissenters have argued that "ecology needs natural history." These outlier scientists believe that ecology requires the systematic study of "organisms in context" to "broaden the scope of science to understand a complex world" in the twenty-first century.[11] The return of natural history, they predict, will be the catalyst that transforms ecological science. If this is true for ecology, why not primate science?

I managed to free myself from my quantitative straightjacket when I realized how much scientific work natural history did by documenting the *context* and *chance* and *contingency* embedded in a group under study. It captured *complexity*, broadly defined, across scales of time and space. This is why no matter how many quantitative studies I did, I still needed natural history to make meaningful sense of the numbers. How could one be expected to paint the Mona Lisa using just two colors? I needed the entire palette for my portrait of baboon society.

To be sure, I'm not suggesting we replace quantitative studies with Darwin's natural history. But I see quite clearly the advantage of a hybrid approach. Natural history has the singular ability to create a whole from the scattered bits of information produced by hypothesis testing. I'm simply saying that natural history should be added to our scientific toolbox—and be respected.

Natural history implies "stories." And the fact is, there's an element of narrative in any scientific paper. The essence of science, after all, is the quest for discovery, perhaps the most basic of storylines. But the telling of stories, aside from a clinical account of hoped-for findings, also got a bad reputation thanks to the "postmodern" critique of science advanced by science studies. In the 1980s and 1990s, this camp argued that all science is "just stories," no better or worse than other stories.[12] I like other aspects of science studies, particularly its focus on reflexivity—asking scientists to think about how they do science—and scientific practice. But I strongly disagree with its stance that all stories have equal weight, making science just another hegemonic master narrative. Stories built on opinion, unsubstantiated speculation, or myths are undeniably different from scientific stories based on facts. But facts alone, detached from explanatory context—a story—lack meaning and impact. Today, scientists toggle between two contradictory constraints. Telling "stories" is unscientific, yet a scientist must tell at least a rudimentary story if factual findings are to be understood and seen as mattering.

As humans we like, need, and want stories because stories and analogies help us think.[13] Stories help us understand the world and perceive the relatedness of things. Though I haven't retained technical minutia from *The Mermaid's Tale*, the book that changed my way of thinking about evolution, a phrase stays with me: It takes more than yeast to make bread.[14] An apt analogy can serve as a surrogate for reams of words, facts, and information. What a wonderful way to think about moving from a one-dimensional timescale to real life: To bake a cake you need more than flour, sugar, eggs, and a leavening agent. You also need pots, pans, a stove, mixing bowls and maybe even a timer.[15] I might not remember the exact relationship between cells, genes, and environment, but I remember this "story," a very short one, about how to bake a cake. It speaks to me of the complex processes involved in real life.

I believe natural history has always been necessary, offering as it does a way to make sense of the relentless churn and change in our lives and all around us. It helped me see complexity in baboon society, but today complexity is everywhere: in habitats, ecosystems, landscapes, and entire biomes. Rapid change is the inescapable norm in the Anthropocene world. We call it the Human Age because science can no longer ignore the fact that humans have replaced nature as the major evolutionary force.[16] This changes everything. We desperately need to understand the layered and connected processes unfolding in this modern, accelerated ecological timescale. Only when we do can we hope to steer ourselves toward the outcomes we wish to see.

---

I think of Darwin and baboons and realize my research solves his implicit question: *why* an "old male should risk his life to save his young comrade from a crowd of astonished dogs."[17] I found the answer: The old male and his young comrade were friends, companions who spent time together, groomed each other, and formed a bond that entailed a tacit guarantee of mutual support. I have seen male baboons risk and even lose their lives while defending friends from predators. I think Darwin would like my answer, and also how I reached it. As for me, Darwin helped resolve my conundrum about methods and the true nature of science. Indeed, I have adopted Darwin as my agonistic buffer when critics challenge my baboon natural history. I have stopped being defensive about baboon stories because I know good field science needs systematic and comparative natural history, as well as hard quantitative facts.

Chapter 17

# Science in the Wild

**A BABOON GENERATION** averages seven years. During my time in the field I have known at least seven generations, the females most intimately because they're always there. But when a new male sticks around for five years, or when you watch a silly scamp grow into a mighty male, they too become known and often loved as well. Friends have asked how I can stand by and do nothing when one of these baboons is sick or injured. The answer is, because I am a scientist. I did break the fieldwork code of nonintervention one notable time. Yes, the translocation. But the scientific promise of the translocation motivated me as strongly as the need to remedy an intractable situation, harmful to both baboons and people. That meant my way of doing things, for a long while, found me on the outskirts of science. But aside from my attempt to save Peggy, I've made few exceptions. I admit to chasing the occasional dog, and I still ask local folks to discourage baboon visits. Meanwhile, lessons are learned and research marches on. In this chapter, I offer a kaleidoscopic view of what it's like to do science in the wild.

I have given much thought to the moral, ethical, and practical issues that inevitably arise when you study creatures in the wild whose lives would be longer and healthier if you'd chosen to make them so. Especially when they're close biological cousins, whose ways so often seem eerily similar to ours. Moreover, how do you follow three and sometimes more baboon troops who travel many foraging miles each day and may go places where a fused back is a distinct liability? Who will watch the baboons when you can't be there? You have a team, and their safety is a consideration, aside from concerns about consistent data collection. It's the management side of science in the wild, and it too comes with rewards as well as challenges.

There are no simple answers to any of these issues, but my scientific goals help with clarity. The point of my studies has been to discover patterns of behavior that tell us both about *what* baboons do and *why*. If, for example, I decided to break up fights, play with youngsters, or medicate the injured, I would have to question the authenticity of the behaviors I saw. I have wondered if the team and I hovering around the troops for so many years might have caused a subtle loss of naturalness. Was it always "normal" behavior we were seeing or possibly something else, perhaps the result of cumulative micro-interventions over decades or a loss of healthy wariness of humans. In any case, after moving the baboons, I doubled down on my noninterference policy because the study was now a test of baboon survival and adaptability. We needed to know if translocation would work as a conservation and management strategy, not just for baboons but for other primates as well.

I fully subscribe to the critical importance of following standard rules about field research. Scientific rigor is essential, and credibility is at stake. But rules don't erase feelings, and at times I've felt torn in two. I have learned that as losses accumulate, grief becomes a sort of shadow companion. Thoughts of Peggy's death or Wiggle's can still bring tears to my eyes. I have found that grief is assuaged more effectively by distance than time. When I've returned to California and had classes to deal with, I don't dwell. But back with the baboons, I'm faced with reminders.

---

The death that affected me most deeply, after Peggy's, was that of Zilla. She was the last of the old females who'd come over from Kekopey, and the top female in Malaika. Like Peggy, Zilla was special, possessed of baboon charisma, at least for me. She was smart, social, and held her own when challenged. She felt almost like a friend, if friendship can be unrequited. She was 22 when she died, quite a bit younger than Peggy, who was at least 32 when she fell from the cliffs at Kekopey. But life on Chololo was harder, and females aged quickly.

In 1997, during a drought that followed several dry years, Zilla was clearly suffering. Her eyes were sunken, her hair unkempt, her manner listless. One evening in early March, she was missing as the rest of the troop began their climb up the sleeping rocks. For three days we didn't see her. The troop made do without Zilla's almost preternatural ability to find the best available food that others had missed—acacia seeds or the first green flush of grass on an old boma site. When food was prolific, her knowledge mattered less, but in times of scarcity it was crucial, when locating a lone *Acacia etbaica* tree in flower could be lifesaving.

We assumed Zilla had died, and I felt a terrible heaviness. On day four, the team and I walked with the baboons to a gully where a temporary human encampment now sat. There, shockingly, was Zilla, but not the Zilla of four days ago. She was even more gaunt, and her face was crisscrossed with dried blood. Matted hair at both temples partially covered nasty-looking injuries. Her lower face, around the jaw and chin, was swollen and disfigured. I saw traces of maize on her lips and figured her cheek pouches were full. Most likely she'd been attacked by dogs after finding stored maize at the boma.

As we approached, Zilla turned to leave the boma. She made no effort to rejoin her troop. But as she rose, she lost her footing and fell. She pulled herself up and took a few halting steps, looking like the proverbial drunken sailor. There were no visible injuries on her legs and hips, but her weakness told me she was probably close to death.

I asked the team to stay with the troop as it moved around the boma and away from the barking dogs. I followed Zilla, glad to be on my own. I'd had trouble concealing my distress at the sight of this wonderful female in such dire shape. Even writing these words is difficult. I quietly trailed Zilla as she staggered a hundred feet or so to a tree where she sat and rested in the shade. She roused herself up once more, lurched forward for a few dozen feet, then stopped and rested again. At this point, I cracked, no longer able to hold back tears, my heart broken by her sad, valiant determination to keep moving. And yet . . . the scientist in me wondered why she kept going. She seemed intent on getting somewhere, as if aiming for a particular destination.

During her brief rest breaks, Zilla worked to chew the maize still in her mouth. Normally, crunching the hard, dry kernels into meal would be easy. Now, with her jaw so swollen, the kernels fell from her lips despite her efforts to push them back in. Then, once again, she'd hoist herself up and make her way forward, always with seeming intention. For nearly an hour I watched the painful process, wanting desperately to rush to her aid, yet knowing I should not. Something was happening here, and I had a hunch about what it was.

I followed Zilla as she struggled up a small rise. If she was aware of my presence, she gave no sign. As we crested the hill, I saw her target, a rare drought season sight on Chololo: a large, flowering acacia tree. In this food desert, it loomed like a mirage. I did not know this tree was there, nor was it visible at any point during Zilla's tortuous journey. But Zilla had been aiming for it, and she got there by dead reckoning. Her body was failing, but her food-finding genius was intact.

But now she faced her greatest challenge. The acacia flowers bloomed at the tip of thin branches that, standing on tiptoe, I could not reach. Getting to those flowers would be no problem for a healthy baboon, but Zilla was far from that. Frail as she was, she managed to get into the tree and pluck a few blossoms, but she nearly fell more times than I care to count. I stayed with her until late in the day, when I had to get back to camp. The next day, I returned to the tree, but she wasn't there. I never saw her again. Her last gift to me was a scientifically valuable depiction of baboon intelligence and grit.

---

As a woman doing fieldwork in a "dangerous" place, I felt strangely insulated from feelings of fear, despite my introduction to the baboons by a man who clearly thought he'd narrowly escaped death by warthog. To my mind, I was simply doing what needed to be done. Then, as time passed, I found courage from the baboons. Not incidentally, they provided an excellent warning system with their natural vigilance and chorus of "wahoos" when danger was sensed. Later, they gave me courage about what I'd learned from them. I had made sense of their behaviors, even ones I'd never seen, because, over decades, I had cracked their code.

I took surprising intellectual and physical risks. In those early days, I left the van even though I was yet to learn that male baboons were anything other than the massive monkeys described by the baboon model—perhaps thinking innocent until proven guilty. Then I named the baboons, which broke the scientific taboo against anthropomorphism. Of course, I wasn't the first. Jane Goodall named her famous chimpanzees, then was roundly criticized for being unscientific. Goodall wasn't my inspiration, however. I knew

I couldn't find the answers I needed if I didn't know each individual. But what was the point of naming baboons if I couldn't track them closely? In fact, I was part of an important shift in scientific practices, but it took time.

I grappled with emotions that I knew a man in my position may not have admitted they felt then, such as the first time I saw a male baboon catch and eat a baby Thomson's gazelle (see chapter 4). At that time I was tape recording my notes. Later, when I transcribed them, I heard sobs punctuating dispassionate description. Though I got better at compartmentalizing, my emotional bond with the baboons grew stronger over time. I worried about that. Would my emotions invalidate my science? But the crop raiding research and the translocation convinced me otherwise. I could care and I could do good science, as long as I kept the two separate. Today, caring about your subjects is no longer taboo, provided that the scientific perspective prevails.

---

There were times when no one felt safe, and the question became one of science versus personal risk. That the baboons proved able bodyguards was a godsend. Ever alert to a predator's approach and raising the alarm when one did, the baboons were allies, especially when they ranged into the new wildlife sanctuary. Protected from human hunters, prey animals multiplied and attracted savanna predators we'd rarely seen in our daily walkabouts: lions, hyenas, the occasional cheetah, and most importantly for the baboons, the leopard.

One day I arrived at White Rocks before dawn and found Malaika already at the base, clearly disturbed and anxiously barking: "Wahoo! Wahoo! Wahoo!"* I poked around the area and noticed fresh

*Baboons also wahoo when they're lost or left behind, but it's a longer, less intense vocalization. The warning bark is a sharp staccato.

blood, then discovered a bloody trail leading from the lowest of the sleeping rocks to a thicket. Telltale fresh leopard footprints marked the ground alongside; a body had been dragged. I did a quick census. An adult female was missing. I had heard her infant moaning and making piteous wahoo-hiccup sounds but I'd mistook them for weaning distress. The leopard was gone so I was safe. But now I saw clearly why baboons compete for the highest sleeping rocks. While all three White Rocks outcrops provided a degree of protection, the lowest one in the cluster was the least safe. This is where the troop was forced to sleep last night when Soit occupied the best two rocks. A leopard had been bold enough to take a baboon in the dawning light.

The lion encounter was more directly threatening. Two research assistants and I were tracking a Malaika consort group. The consort pair and a clutch of male followers had made its way into a thicket west of White Rocks, ahead of the others. It was an unusually turbulent group, with its large crew of determined follower males and a fretful consort male. My attention was fully trained on them when, like thunderbolts out of nowhere, two female lions charged. Typically, our human presence would have deterred a lion attack on baboons. Maybe the lions had been unaware of us—unlikely—or maybe they were so ravenous they didn't care. In any case, it happened so suddenly that fright didn't register, at least not initially. The males in the consort group turned away from the female and faced the lions, pant-grunting defiantly, while we three humans screamed and hollered. The lions turned and left, though I had no idea if they lurked nearby or had gone a distance. Only then did I notice my physiological response to the incident. My heart was racing and my knees felt like jelly. After a few moments, I collected myself and tracked down the consort group, back in business as if nothing had happened.

Luckily, no one got eaten that day. However, a few days later, I spotted a group of lions sunning themselves not far from where the baboons were headed. I turned back, but the baboons kept going. The next day, two female baboons were missing. Malaika and Soit would go on to lose many animals. Indeed, predators brought the two troops together, because, as we learned, there is safety in numbers (chapter 9).

---

Risk comes in many forms, and for me it involved limitations imposed by my back. One time stands out. I was out early with Pumphouse, shivering as usual in the crisp morning air. Pumphouse had taken to sleeping at a higher altitude, on a ridge close to the Mukogodo Forest, a remnant dry forest. My fixed back could manage the drive there over 13 miles of rutted road, though not comfortably. After a morning of baboon watching, I headed back down the ridge toward the car. I lost my footing on the steep slope and grabbed a tree branch to steady myself, then heard a snap as it broke off in my hand. Gravity took over. I slid on my back halfway down the slope, where, fortunately and unfortunately, a *stricta* cactus plant stopped my descent. Wedged between the pads, thorns, hairy fruit, and the ground, I checked myself. Nothing was broken, but I had egg-sized swellings on my chin and head. My fused back meant I couldn't roll or shimmy my way out, as I once might have. Inch by inch, I scooted away from the plant, my jacket and slacks bristling with cactus thorns and glochids. I knew when I shared the mishap with Jonah, he'd say, for the umpteenth time: You have limits, *please* accept them. He did, and I changed my ways. The fact is, I was lucky to escape fresh damage to my jerry-rigged spine, or worse. I now carry a hiking stick in the field and find detours around slopes, though it takes more time.

---

After the translocation, I had laid down an ironclad "no interference" rule. Otherwise, the scientific value of the move would be void. If I intervened each time a baboon got hurt, sick, or stressed, I'd be interfering with baboon adaptation and delegitimizing my study. Worse still, I would make these animals wards of humans at the expense of their self-determination and independence. As it turned out, the loss rate in the troops I moved was lower than at Kekopey, where crop raiding led to many deaths. In fact, over time the translocated baboons did as well as the indigenous troops we'd added to the study.

Thankfully, the baboons do a pretty good job of taking care of each other. But not always. Most typically, the sick or injured lag behind, often far behind. Unless you stay with the disabled animal—difficult when you have a whole troop to study—you're unlikely to see it again. As we saw in chapters 9 and 13, a vital troop function is safety for most, if not all. Separation from the troop is risky for an injured baboon *or* an animal inclined to help it. A baboon on its own for any reason is at risk, because the troop keeps moving in its search for food. Admittedly, several species do better than baboons at caring for each other, notably elephants, zebras, and dolphins. All have been observed hovering protectively around injured members of their groups. But in baboons' defense, those animals are larger and better able to fend off would-be predators.*

But I have seen more than a few incidents that convince me of baboons' capacity to care. One involved Roger, the infant son of Riva, a descendant of Robin. Little Roger, who, at five months, had just started to turn from black to brown, was badly injured with a gash across his body. (There was no way of knowing how it happened.) When the troop lit out from White Rocks, Roger couldn't or wouldn't follow. I stayed with him, sitting nearby, but of course not engaging. About ten minutes later, a favorite juvenile playmate

*These more protective species also have smaller groups with fewer foraging challenges than those of baboons. This helps free them to protect their own.

appeared. First, she groomed him, then gently coaxed him back to the troop, which was still not far from the rocks. The scene was almost unspeakably touching. Roger survived his injury only to be taken by a leopard a year later. But his playmate had given him that extra year.

In another incident, Okello, a juvenile male, appeared one morning with a severe wound on his callosities. Again, we could only guess the cause. The troop moved on from the sleeping site, leaving Okello on his own. His screams and wahoos brought back an aunt and cousin. After sitting with him for a bit, his cousin took his hand and tried to pull him to his feet, meaning to lead him back to the troop. But his wound was too grievous. The two females finally rejoined the troop, leaving Okello behind. I held out hope that we would find him still at the base of the sleeping site when we returned in the late afternoon. But he was gone, no doubt taken by a predator. Again, the safety in numbers lesson was vividly and sadly clear.

But few episodes compared to Squashy's sacrifice. Leopards regularly kill baboons (and livestock and dogs) in the cover of night, so we rarely saw those dreaded predators. But exceptions did occur. In one such case, Pumphouse had just left their sleeping rocks when we heard screams, leopard growls, and baboon alarm calls. A leopard had attacked a female and her infant. Squashy, their male friend, swooped down to their rescue. The mother and baby got away, but poor Squashy wasn't so lucky. The leopard fled as we got close, but it had already managed to stash Squashy's body in a tree. We lowered his remains before the cat returned. Squashy's bones now sit near those of Peggy, a poignant reminder of the power of friendship. Add to these anecdotes many involving baboon mothers who hold and "care for" a dead infant for several days, unwilling to break the bond. And then, of course, I think of Darwin's old monkey who saved his young friend from a snarling dog pack, an iconic baboon rescue.

Behaviors like these resonate with our human emotions. They're familiar as well as poignant, and they remind us that the differences between our species may be fewer than the similarities. No wonder we're tempted to think of them as different versions of us. It's an issue in the field, and in the following chapter I delve into the subject of anthropomorphism.

---

Tracking baboons is harder than it looks, demanding as it does undivided attention and focus on subjects that often seem to move so fast that it is hard to keep up. A half day spent collecting data seems the human limit. I often watched in the morning or afternoon, while research assistants tracked the entire day in half-day shifts. We were in this for the long haul, unlike the more typical graduate student shorter field study. Thorough, dependable coverage was crucial.

By 1976, I realized I had a "project" and that it was scientifically important to keep it going. I needed help in the field, especially as more baboons came under study. Not only did baboons need tracking, but continuity had to be assured during my annual teaching quarters in California. There was no shortage of graduate students wanting to cut their teeth on fieldwork. The Gilgil Baboon Project was formed and, once official, began recruiting grad students. But as crop raiding worsened, I understood the necessity of working with the community. The data-focused students were not inclined. So, in the early 1980s, I changed course, phasing in Kenyan research assistants who viewed the work as a job and knew the language and local customs—all great advantages. Most hadn't completed high school but were easily trained. After the move to Chololo in 1984, we became (and still are) the Uaso Ngiro Baboon Project. The Kenyan research assistants and I shared a ramshackle house, the one that reminded me of the Australian outback with its rusty corrugated iron roof, mud brick interior walls, and rough stone exterior. I am by temperament an egalitarian, but the Kenyans viewed me as "boss," a title I hated. Team members had bridged the cultural divide when they accepted a job studying baboons, something their family and friends thought was absurd. But that didn't mean we shared cultural perspectives, though we tried. I wanted staff to comfortably unwind after hours, so we set up a rustic wing for "the boss."

I never mastered Swahili, but I learned enough to get by and to appreciate its charms. For example, *chafuka* is Swahili for making a mess. Team members transformed it into "*chafuka*-ing," which happened when one of them tried to master the data filing system. Other neologisms are similarly droll. The word for bicycle is *baiskeli*, a transliteration from one language into another, much like *sitema* for steam or *rongo* for a lie. Swahili often makes use of onomatopoeia, as in the words for motorcycle, *piki piki*, and generator, *tinga tinga*. Many new words are wonderfully creative. Thus, the prefix "wa" (referring to the plural) becomes Wabenzi (people of the Mercedes Benz) to denote a wealthy person.

My favorite linguistic encounter happened one day in the field. Pumphouse was at Windmill Gully, and for the first time, vervet monkeys showed up. Much smaller than baboons, vervets have black faces and long tails. Lawrence, an assistant who started with me at Kekopey, was the first to spot them and ran to alert me. Initially, I had no idea what he was saying. Finally, I figured it out. He had seen "baboon puppets," an apt and priceless description for vervets.

---

I love going out to the baboons just to watch. Some days I set aside my clipboard and datasheets and simply look around. In these quiet moments, my romantic self overrules the scientist and nature feels magical. But these reveries also lift the curtain on the big picture and allow me to see connections missed while tallying behaviors. Today I watch Gama, a favorite. He is one of Mavis's sons who never left the troop, perhaps because troop fusions provided unrelated females. Gama grew into a powerful large subadult male who, following the male trajectory, became dominant. Eleven years later, he is no longer dominant but still popular, always surrounded by friends of all ages, including a female who grooms him. Close by, youngsters tease and frolic. This classic little subgroup stays together as the troop leaves the rocks. Infants and juveniles often orbit their male friends throughout the day.

Contemplative baboon watching is not unlike Darwin's famous backyard strolls, a practice conducive to finding threads between seemingly unrelated things.[1] It's an unmatched perk of doing science in the wild. Once in a while, when I sensed the presence of an open mind, I dared to admit that I'd begun to see the world from the baboon's point of view. But it was not the sort of thing one often—or, really, ever—heard coming from the mouths of "serious" scientists, even if my inspiration came from no less a scientist than Charlies Darwin. While measuring, logging, and classifying his observations as the *Beagle*'s naturalist, Darwin trained his mind to "see" what was around him.[2] I did the same, only I went further when I put on my baboon glasses.

Chapter 18

# Interpretations

BEAUTY MAY BE IN THE EYE of the beholder, but so, in a sense, is science. I have already touched on shifting styles of science, but here I'd like to delve more deeply into issues of method, perception, and bias, including gender. They all bear on the interpretation of scientific findings, including mine. Can the objective ever be free of the subjective? Perceptions and interpretations matter because they filter our sense of how things work, shape perspectives, and incite actions. As to nonhuman primate behavior, interpretations have consequences because they inform our ideas about where we came from and who we are. This explains the keen interest shared by lay and scientific communities alike in discoveries about our distant ape and monkey cousins.

Let's begin with words, their meanings and shadings. The words we use to describe animal behavior obviously come from our human language, with its precise descriptors for human behavior. Language provides a direct route to anthropomorphism—projecting human behavior onto animals—the classic bugaboo of any animal scientist. Our inclination to endow animals with human qualities raises issues around both semantics and interpretations. I found it difficult, and also inconvenient, *not* to use the words we use for our own behavior to describe what baboons do, and it's gotten me in trouble. As has my inductive approach, in which I set aside all presuppositions and allow the baboons to guide me toward understanding. But how to label baboon behavior is a lingering problem, and not just for me. Our linked biological ancestry makes it harder yet, because they are so like us, yet not, and what are we supposed to make of that?

In Darwin's wake, most studies were glaringly rife with human bias as investigators tried to reconcile the newly "discovered" closeness between humans and their nonhuman primate counterparts. Then came the pioneers of modern animal studies in the 1930s, who fiercely rejected this anthropomorphism. These classical "ethologists"* corrected for anthropomorphism with methods and language that scrupulously avoided the "taint" of human qualities. In this scheme, animals had NO intentions, NO emotions, NO mind, and absolutely NO personality. Descriptions used clinical terms like "fixed action patterns," "imprinting," "releasers," and "instinct." Behaviors such as predation, communication, and mating were tallied in units using a format called an ethogram. I used one in my first study. Once recorded, these "objective" units were reassembled into sequences. Thus, a duckling, following *instinct*, *imprints* on its mother shortly after birth because it needs to follow the "mother-figure." Konrad Lorenz famously used this causal analysis to show ducklings imprinting on him—following him—absent their mother.

I illustrate this approach to my students by walking across the classroom to the door. A casual observer would say, "The professor

*Those who study animal behavior from a biological and evolutionary perspective.

is leaving the classroom," a statement suggesting human intention. The ethological approach would break my movements into "fixed action patterns" that describe how I move my leg, foot, and body, one step after another, until I reached the door. No meaning, no intention, no human biases.

These early ethologists expected that once pieced together, fixed action patterns would yield the bias-free *meaning* of the behaviors they'd recorded. They successfully identified those patterns, but the meaning part proved stubbornly elusive. So they tacked instead to a long and spirited debate about nature versus nurture. Now that controversy has given way to a more nuanced consideration of genetics, epigenetics, physiology, and other comingling factors that modify both nature and nurture.

I was trained in this ethological tradition, and soon I was hitting my head against a wall. The tools I was given were not serving me well in the field. The simplistic, atomistic ethological language seemed to squeeze the very life out of behaviors, interactions, and patterns I saw every day. Put another way, this limiting language proved an enemy of accuracy in my descriptions of baboon behavior. Imagine being asked to analyze *War and Peace* using a six-year-old's vocabulary. It was like that, and it left me at a loss for "words."

For many years, the conundrum of anthropomorphism complicated my work. If it was unkosher to describe baboon behavior using human terms, then what were the options? I dug into the meaning of words that describe human behaviors, figuring I could justify applying them to baboons if, technically, the baboon and human versions were close to identical. I grew comfortable using the term "negotiation" to describe baboon interactions, which so closely resemble those of humans. There simply was no better word. Next came "friendship," which replaced Tim Ransom's neutral term, "special relationship." This adoption took time. But the data and my observations finally convinced me that baboon friendships matched those of humans in key respects. Later, using the same criterion, I added "collaboration" and "trust."

Early on, the problem became more than conceptual. With my first major paper about the baboons' new style of predation on young gazelles, I aimed to wow the reviewers with my scientific rigor. The human word "hunting" carries a lot of cognitive baggage. When humans hunt, they often do so cooperatively, based on a plan. I wasn't certain that baboons hunting in a group met human criteria. When I wrote up my findings, I used "neutral" terms, labeling human hunting as "p hunting" and baboon hunting as "q hunting." But when I submitted the paper—a version of my doctoral thesis—the reviewers strongly objected. It seems they thought I'd gone overboard in my effort to avoid anthropomorphism—ironically, given how things turned out. They insisted I use "hunting" without qualifications. That was in 1975.

Then came my paper about the male hierarchy.[1] The reviewers, all male, rejected my findings that the male hierarchy was fluid and that rank did not correlate with reproductive success. But publication of the paper hinged on my first supporting the male dominance hierarchy. To comply, I devised a weak criterion: A male achieved dominance if he won 51% of his encounters with other males. It was, perhaps, a shifty way of illustrating the fragility of the male hierarchy, since the victors in those encounters changed regularly. (Later, my conclusions were confirmed by other investigators.)[2] The reviewers still grumbled, but I got the paper published without irreparably compromising my stance. That was 1982.

Then, in 1987, my first book, *Almost Human*, came out. My early revisionist ideas about baboons were on display, in a style targeting both mainstream readers and scientists. The reception was, predictably, mixed. Lay readers and scientists in other fields loved the book, colleagues in my field . . . not so much. They were skeptical of my interpretations—that baboons had friendships and often used these in social strategies to avoid conflict—and few bought my debunking of the male dominance hierarchy. Still, it was a Book of the Month Club selection and sold well.

These weren't the only dustups I've had with the scientific establishment, but they amply illustrate why I have spent my career careening between confidence and self-doubt. Suffice it to say, I moved from inside the scientific paradigm to outside of it. Thankfully, bravery is a benefit of getting old. (In my seventies, I no longer can fool myself that I'm just middle aged.) I don't much care about what others think and have little to lose by expressing my ideas.

Lately, I've been heartened by long overdue changes in scientific attitudes, reflecting the broad shift in social values and attitudes over the past half century. Some animal behaviorists will now admit that nonhuman primates have mind, emotions, and personality, and even that the older methods were problematic. In fact, we're at a juncture where a few vocal scientists openly reject an anti-anthropomorphic bias *because* it robs animals of abilities and skills and ignores their emotions and mind. The late primatologist Frans de Waal, who wrote bestsellers about chimps, dismissed the old scientific approach as "human exceptionalism."[3] Still, there is a space between "almost human" and "little humans." I continue to ask myself a simple question with boundless implications: What does it mean to share so much and not be human?

Adopting a baboon point of view gave me novel ideas about baboons that still get a mixed reception. Take the fusion of Malaika and Soit. The merger of two separate hierarchies required fine-tuned social engineering. My observations yielded strong evidence supporting my contention that a baboon social group is negotiated. Since some don't see it that way, it becomes a matter of interpretation.

---

Adopting a subject's point of view isn't as crazy as it sounds. Barbara McClintock was the first woman to win an unshared Nobel

Prize in Physiology or Medicine, for her investigation of chromosomal changes in maize during reproduction. Later, she said she had developed "a feeling for the organism," by which she meant she'd "trained" herself to "identify" with her subject—maize.[4] More recently, Hope Jahren, in her book *Lab Girl*, talks about her scientific about-face, when she decided to study plants from their point of view to "puzzle out how they work." These were brave admissions coming from brave women. Despite undeniably brilliant results, McClintock's approach was widely criticized at the time. Jahren braced herself for a similar reaction: "What was there to stop me, aside from my own fear of being 'unscientific'? I knew that if I told people I was studying 'what it's like to be a plant,' some would dismiss me as a joke."[5] Learning that these scientists had found ways to inhabit plants made me think I had it easy with baboons.

Still, I had to wonder. Was I able to understand the things I saw because I was a scientist, because I was a woman, or because I was a woman scientist? From the start, I did science differently, but I'm not sure if the difference was idiosyncratic or gendered. Did I test the baboon model with new methods because I was a woman? Probably not; I didn't just record pair-wise encounters between baboons—the norm—but added in social context. Then I grew the context to include multiple troops and ecological variables, as I unknowingly dipped into natural history. New discoveries resulted. I suspect the term "maverick scientist" applies as accurately as "feminist" or "woman scientist." Just *scientist* would be my preference.

Many women scientists I knew were mavericks. For example, Thelma Rowell was a true groundbreaker in the field of animal science. Thelma retired to Yorkshire to raise sheep and, being Thelma, she watched their behavior. Sheep, she concluded, had much more primate-like social relationships than anyone imagined. Did Thelma recognize the "primate-ness" of sheep because she had studied monkeys, because she had an open mind, because she was a woman—or all of those? I'll have to ask her.

Yet I did see things my male predecessors missed. Indeed, I found myself an "icon" of feminist science after news outlets picked up on my claim that female baboons anchored the group and had their own hierarchy. Still, I never took sole credit for the finding. However, my study was the only one framed as a test of the male-centric baboon model. In time, the hoopla died down as more women primate scientists focused their research on females.[6] The status of female primates was thoroughly researched.* In any case, feminists embraced my work as evidence that women do science differently, in a good way.[7]

Despite my seeming ambivalence about the feminist label, please know I marched for women's rights and supported the feminist agenda in my Berkeley days. But I didn't identify as a radical feminist and my research had no hidden feminist agenda, at least not consciously. In graduate school, I was aware of the unusual status of "woman scientist" and looked around for role models. Of course, in anthropology there was no shortage of great women, including Margaret Mead, Ruth Benedict, and Elizabeth Colson. By the 1960s, Jane Goodall, Thelma Rowell, Alison Jolly, and Phyllis Jay had already studied primates in the wild, so I would be in good company. It was harder to find a woman who successfully combined a career and family.

I started out giving equal time to large males and females—and glimpsed the female hierarchy—but the male system captured my attention for the next decade. It was after the translocation, when the Pumphouse baboons had a spell of turbulence, that I clearly saw

*Recent popular books have rediscovered and found significance in these findings about the important role females play in nonhuman primate societies (see *Bitch* by Lucy Cooke and *Eve* by Cat Bohannon).

the critical stabilizing function of the female hierarchy (chapter 10). And, yes, there were gendered issues in animal science, most especially in the study of nonhuman primates.[8] I deliberately avoided research into mother-infant relations and infant socialization, both considered "women's" topics. In fact, I may have overcompensated on the side of gender neutrality after that first study. I set aside my early impressions of the female hierarchy and focused instead on the social tactics devised by male baboons to avoid aggression. This was an unorthodox idea at the time, and perhaps more exciting than orderly females.

Still, for a long while there was no getting around the regrettable "specialness" of being a woman in science. There were times I questioned if I was truly a "scientist." The answer varied, depending on temporal definitions, both mine and those of my colleagues. Recall that in the early 1970s, Science (then with a capital S) was "hard" science, apprehending reality through experiments, quantitative data, mathematical modeling, and statistics—what came to be called "the physics envy" definition of science (see chapter 16). I sometimes felt excluded from Science because anthropology gets lumped in with the so-called soft or social sciences, à la sociology, psychology, economics, and political science. That was why I was so intent, early in my career, on striving for a new and unimpeachable objectivity.

---

The question of my scientific bona fides became moot when I finally recognized the distinction between Science (capital *S*) and science (small *s*).* Traditionally, science was seen this way, as a generator of absolute and unchanging Truths. But this aura of infalli-

*This was a major intellectual shift made possible by my collaboration with Bruno Latour and his investigations, which restarted Science Studies as a field.

bility was a cultural construct, similar to the traditional images of all-knowing doctors and the holy church. Small *s* science, on the other hand, is simply research that allows facts to change as methods improve and new information emerges. Because "Science" was seen as the last word on any matter it investigated, the public was bound to lose faith when "science" began to question those Truths. Today, trust in science mingles with skepticism. What we need to convey is that research is a dynamic and ongoing process that aims to clarify our vision of reality and adjust it as needed.

Be assured: I firmly believe in the power of the special tools used by science. They've taken us into space and into cells, helped us create vaccines and the internet, and given us insights about consciousness and our origins. However, as I learned from my detour into science studies, it is a myth to think that science can be isolated from society because facts can't speak for themselves.[9] Mere mortals circulate facts and thus make them real. Science studies, which got traction in the 1980s, confirmed what Washburn claimed to me years before: "Facts are the best fit between today's methods and reality; as methods improve, facts will change." Between Washburn and science studies, I had a definition of science that worked for me. I stopped worrying and unapologetically became a "scientist."

---

I convened my first Wenner-Gren international symposium in 1976, calling it "Baboon Models and Muddles." My goal was to engage a select set of colleagues in a discussion of how and why we choose certain questions to study and why we find some explanations more satisfying than others. I needed a participant who could help situate primate science in the context of other scientific endeavors; Bruno Latour was the obvious choice. To my dismay, the meeting quickly went south. My colleagues were not prepared to reflect on their science. Perhaps more surprising was their resistance to

Bruno's plan to observe their behavior in the workshop. After all, they were professional primate watchers. I quashed the move to expel Bruno, but we never got to the key questions and no publications resulted.

That didn't stop me from wondering why some answers, stories, or metaphors are more pleasing than others. I also wanted to know how ideas about primate society might have changed. These questions became topics at my second Wenner-Gren symposium, held in 1996 in Teresópolis, a mountain retreat high above Rio de Janeiro. Given the debacle at the first Wenner-Gren, I was relieved when Bruno agreed to attend. Attitudes had relaxed by then. It was a truly international small group—half men, half women—representing multiple disciplines related to primate science and science studies. We considered why some facts are believed while others are ignored or even disbelieved. The acceptance or rejection of a fact, we decided, involves any number of variables, including its historical context, culture, gender, the fact's source, and the personal and professional backgrounds of those sending and receiving the fact. Each variable may bias perception of *the putative fact*. Given this stew of influences, what an individual chooses to believe as fact may have little to do with science. The implication for the study of nonhuman primates, our not-so-distant cousins, is that our disquieting commonalities easily engage our emotions and color our perceptions in a way that studies of mice or amoebas don't. We're tempted to project ourselves onto these creatures, which of course invites bias. Happily, the outcome of this 10-day symposium was *Primate Encounters: Models of Science, Gender, and Society*, a provocative and well-received examination of the titular issues and our interactions.[10]

Working with Bruno opened my eyes to the "genealogy" we create of our human behavior.[11] That is, we perch ourselves on a branch of the sprawling primate family tree, then figure that our relatedness to other primates explains our behavior, both good and bad—and that it's all inevitable, ordained by evolution. I have always been wary of this concept, and it makes me careful about my baboon *stories*. Einstein often gets credit for this quote: "If you can't explain it to a six-year-old, you don't understand it." Some say it's apocryphal, but even if so, it's a fair point. If I were to explain baboons to a six-year-old, I'd use everyday language, no doubt laden with anthropomorphisms meant to illuminate and accomplish the task with fewer words. Stories that contain relatable analogies create understanding and a constellation of connections. Every chapter in this book contains stories. Most spring from data and analysis, but some are speculative. These stories function in two ways. First, they serve to illustrate baboon complexity. Second, they're meant to hook the reader by describing baboon episodes that, I hope, will linger in the mind. But these stories aren't meant to suggest that we're the same, just similar. Metaphors like Darwin's "tree of life" or Dawkins's "selfish gene" powerfully trigger ideas about relationships and are not easily forgotten.

---

But, of course, the holy grail of interpretations is evolution. While my peers pursued branches of fieldwork designed to test predictions of evolutionary theory, I continued with my quantitative field studies, augmented with natural history. As our paths diverged, mine took me deeper into the nuanced complexities of baboon society. Parts of what I observed fit into the prevailing evolutionary framework. For example, baboons often followed evolutionary foraging predictions. But much did not. I had seen unexpected behaviors, chance events, and baboons facing and choosing different options. I understood the deep entanglement of the social and ecological. My cautiously expressed skepticism came from baboons, not from theories about evolution and science, which made me not just a woman scientist but a contrarian female to boot.

For example, evolutionary theory assumes that kinship is deeply embedded biologically. For baboons, however, "kin" are only "known" through the same process that builds "behavioral kin," defined as unrelated individuals connected by proximity and trust. A newborn baboon infant doesn't "see" her mother and has trouble, for hours, finding the nipple. Snuggled on the warm belly with milk nearby, the baby is forming her first and most important social relationship. Does she "know" that this belly is mom? To my surprise, the answer is no. Females rarely kidnap babies, but one such case was instructive. An adult daughter kidnapped her mother's newborn, her own little sister. Fortunately, the kidnapper had an older baby and was lactating, so the infant didn't starve. A few days later, when big sister wanted to give the baby back to mom, the baby refused. The mother finally got her baby back, but it took some doing; as far as the baby was concerned, her sister, who had held and fed her,* *was* her mother.[12] I concluded that while kinship is technically biological, a baby is born naive and acquires the knowledge of kinship through behavior. Mother's body is ground zero of social awareness. Soon, the baby begins to read mother's reactions to nearby animals and to build out her social network to include biological kin—siblings, aunts, grandmothers—and male friends and playmates. If mother calmly accepts another baboon, then so does baby.

I started out testing baboons as models for human evolution. You may recall from chapter 1 that my mission in college was to find the discipline likeliest to take me to the origins and understanding of human violence. The baboon model seemed to offer a portal. But it wasn't long after I met baboons that my direction changed. Now, after five decades, my mind is changed. I feel that no living species can serve as a model for early humans. Neither baboons, chimps,

*Contact comfort is more powerful than hunger, as Harry Harlow proved with his wire monkey mother.

nor bonobos were the first drafts for human evolution, even though grant money still flows to those looking to prove that apes might have been. In the case of baboons, our lineages diverged some 30 million years ago, but fossils of modern baboons date from two to four million years ago. In this vastness of time, both ancient and modern baboons were uniquely shaped by a history of countless and unknowable accidents and contingencies. Ancient origins and a long and winding evolutionary road make baboons, or any other modern nonhuman primate, to my mind, not productively comparable to early humans, and certainly not to modern humans. As to the behavior we know as violence, I can say with certainty that baboons are far better than humans at devising ways to avoid it. But comparisons are inevitable, so how can we make them useful while avoiding overreach?

Instead of creating "referential"* models, I prefer the idea of modeling "processes" as I find them in baboons. When I observe a new process unfold in baboon society, I wonder if it will help me ask better questions about humans. For example, when I compared baboon hunting and baboon crop raiding, I was able to identify specific elements in each that determined whether the new behavior would become a tradition. Namely, the behavior needs to recur and to be highly visible in order to catch on and stick. That's why crop raiding became a tradition and hunting didn't.

This shifts discussions of human evolution from a focus on *results*—say, material culture or language—to *process* questions about how these innovations might have shaped the social interactions of early humans. In this approach, my speculations about the role of the baboon group as a learning resource for individuals get reframed and applied to early humans (see chapter 14). The question, then, is what changed and how, for early humans, who created ever more elaborate traditions and cultures, while baboons didn't.

*Referential means using a living species as a model for early humans.

Thinking about process leads us in a new direction and raises important questions about *how* we became human. I would like to think the right questions will bring us closer to better answers.

For now, I'm content to lay bare, as best I can, the complexity of baboon lives, and to kindle interest in and perhaps a little respect for these endlessly fascinating animals. Rarely heard are words of praise for baboons, and I'd like to change that. I have tried to understand and interpret the "language of the baboons," not their sounds but how they uniquely "say" things to each other, their social work. And like Thelma Rowell with her sheep, I would like to find out if other animals possess the socio-ecological complexity I've observed in baboons. I bet they do.

Chapter 19

# Forces of Nature

I WAS A CONFIRMED CITY GIRL when I arrived in Kenya, though a quirky one since I'd already spent hundreds of hours staring at caged monkeys. But I was never one for overnight camping trips or campfire singalongs. Of course, I knew from pictures what I would see on the Kenyan savanna, but not what I'd feel once I got there. Those early days spun me around. I was drowning in beauty. The skies were boundless and ever-changing, the landscape lushly Edenic. And the animals. Once I had become de facto baboon, I was just another member of the wildlife family. I came eye to eye with reedbuck, bat-eared foxes, jackals, impalas, and eland. Only the zebras startled when they spotted non-baboon me with the troop. Distant sightings of ornery buffalo suited me fine. I had grown up near the renowned San Diego Zoo, and now it seemed, well, almost pitiful. I became a birder specializing in spectacularly colored species, and the baboons became my "people." It was just us, the ranch owners living at the far end of Kekopey, and a smattering of cattle herders. The environment dictated my work, and I loved it. Hard rain on Kekopey meant I couldn't walk, drive, or watch baboons. I read and wrote during the two rainy seasons, and I learned to parse the clouds and predict rain. I was a life and a world away from Berkeley.

Almost imperceptibly, I was developing a "romantic" idea of *nature* on Kekopey: Nature was pure and pristine, unconnected to humanity's dirty, overbuilt world. I was later to learn that science and American culture were promulgators of this bifurcated model. The US Wilderness Act of 1964 enshrined the concept of nature as a separate, sacred entity, untrammeled and unpeopled.[1] It was this view of nature that compelled me to agree with colleagues about crop-raiding baboons. They had "fallen" from their natural state and were no longer proper study animals. But by then I felt I owed them, and I cared more about them than my standing with peers. Then came the translocation. Together, the baboons and I would start a new life and a new study in a more natural place, Mukogodo.

But people were there as well, though fewer of them. My romantic view of nature meant that I grieved each time the baboons ranged into a more humanized (meaning more people but still few) part of the landscape. I had assumed that the translocation would solve the baboons' "people problem" because arid Mukogodo was poor farmland. But pastoralist families were settling down and their bomas, though dispersed, were chipping away at "wildness." Events were forcing me to question my romantic view of nature.

---

In 2001, I returned from my teaching term at UCSD to find a world transformed. The long rains finally arrived, ending a harrowing two-year drought. Everywhere I looked was green. All the dams were full. Zebras lined up to drink at the water's edge. Usually, it's cold and cloudy this time of year, but on this day the intense blue sky was cloudless. Dozens of giraffes surrounded our research house, the largest herd in years. Zebras brayed nearby and a lion roared in the distance. I took a deep breath and felt renewed, as I always did.

The next morning, I went to see Malaika. In chapter 9, I told how changed they were after the killings. They still clustered at the base of their favorite sleeping site, White Rocks, grooming and resting in the sun, but now numbered only 13. The much larger troop, Soit, was starting its day nearby. To an untrained eye, the baboons might have looked like one big group, but I knew better. Yet instead of pondering the ongoing fusion, I thought about another question, one that has bothered me since the crop raiding on Kekopey, when farmers killed baboons. Who was responsible for the recent Malaika killings? Had habituation eased the baboons' natural fear of humans and made them easy targets? If so, I'm the one responsible. Mukogodo, on the northeastern edge of the Laikipia Plateau, teems with wildlife, but those animals are shy compared to creatures who safely roam protected parks and reserves. I remembered what a Maasai elder told me when we met before the translocation. Resident baboons were known to take goats in the dry season, but they were gone before anyone could kill them. Wild baboons have a healthy fear of people. Had I "tamed" the fear out of the baboons, making them easy marks for humans who wanted them dead? The thought gnawed at me. I considered moving the baboons again, but where? Nowhere would be free of human influence for long. Besides, I was out of energy and out of money. As for research, in a new place I'd once more be starting from scratch. I reached the final stage of grief, acceptance. We had no choice but to carry on in an imperfect, ever-changing world.

---

In 1985, a biology department colleague teamed with Jonah and me to create a new interdisciplinary course at UCSD. Titled Conservation and the Human Predicament, the class addressed conservation issues from a range of perspectives. Thirty-plus years ago, it seemed daringly innovative, and it was a hit. I tackled cross-cultural views of "nature," the evolution of ethics vis-à-vis nature, and media portrayals of conservation. What I learned preparing for this class was no less than revelatory. "Nature," I was surprised to find, is not an objective or scientific concept. In fact, "nature" is among the most complex words in the English language. Its meanings are several, its phenomena are multiple, and its connotations depend on one's perspective. *Nature*, it turns out, is yet another cultural construct.

I had assumed that my feelings about nature sprang "naturally" from an organic reaction to Kekopey's beauty. Not the case. My romanticism was rooted in a modern Western cultural view of nature—a complicated legacy starting with European industrialization.[2] Herewith, a brief history review. Wealth generated by the Industrial Revolution gave rise to the middle class and leisure time. Meanwhile, cities became crowded, unsanitary, and ugly. These colliding trends birthed the Romantic movement, led by a cadre of elite late eighteenth-century artists, poets, and writers who believed that nature was the panacea for society's ills. Immersing oneself in nature would restore balance and correct for the harm done to mind and spirit by city life and heartless industry. Romantics invoked a mythical past when humans and nature lived in harmony, à la Jean-Jacques Rousseau's "noble savage." It was a view that marked a sharp break from the Christian dictum granting humans God-given dominion over nature. But class and wealth allowed Romantics to sweep aside an inconvenient reality: Farmers had good reason to distrust nature, which often and capriciously thwarted their efforts. Scratch the surface of European fairy tales and you see themes of both dominion over and fear of the natural world.

When I dug deeper, I discovered a striking shift in ideas about nature and wildlife, one which had people in the developed and developing worlds changing places. Prior to the Enlightenment, the biblical view of nature as slave to man's needs prevailed.[3] As we humans grew richer, healthier, better educated, better equipped, and more urbanized, we no longer lived in nature and insulated ourselves from its power. It became an "other." Redefined as a separate entity, nature gained, in our minds, the "right" to exist. We invested nature with purity, nobility, and redemptive powers as well as fragility and corruptibility—therefore deserving protection and rights as well as reverence.[4]

Today, much of the "West" views the Serengeti plains in Africa as an iconic natural wilderness, with wildlife abundant and humans sparse. Yet Jonah reminds me that humans constructed this savanna, starting with hominins using fire to hunt game a million years ago. The result was a grassland savanna, burned repeatedly to prevent regrowth of bushes and trees. What followed much later was an ecological dance between pastoralists and elephants.[5] Over the last few centuries, people and elephants avoided each other, yet fashioned habitats that happened to serve the needs of both. One could reasonably call the Serengeti the largest cattle ranch on earth. Elephants graze, cattle graze, and so do some two million wild hoofed animals of varying types. Is the Serengeti a product of waves of utilitarianism or a natural wonder? Depends on your cultural perspective.

I think of a telling incident. A young woman, the daughter of an American friend, visited the Baboon Project. I always ask each visitor to share something with the team. Responses vary wildly, from the professional (research results) to the random (ever wonder why Switzerland has no standing army?). When this young woman spoke of her animal rights work in the United States, the conversation took an interesting turn. The Kenyan research assistants politely pointed out that the rights she wanted for wild animals—for example, the freedom to roam and protection from harm—included rights that they, as humans, still lack in Kenya.

That awkward moment perfectly captured how divergent cultures forge divergent perspectives.

To test whether nature is "cultural," I posed a question to the Project's Maasai pastoralist research assistants. On a particularly beautiful morning, as the sun rose behind White Rocks, I asked Martin what he saw. "Trees," he said. "Not a glorious sunrise that feeds your soul?" "No." I got a similar answer from Francis. Clearly, our cultural lenses shaped our perceptions of the sunrise—me, wonderstruck, my associates, meh. Yet the Maasai researchers unanimously agreed that cows are beautiful. Suddenly animated, they could discuss cows for hours. No further proof was needed of culture's inestimable power to shape perception. Culture is everywhere, an invisible, unconscious, emotionally powerful filter on all human behavior. That means nature can be a sacred forest or a timber concession. Cows can be revered or eaten. Shark fin soup can be a delicacy or a horror.

---

In the distant past, traditional "small-scale" societies were members of the natural context, not outside it. Nature was both benefactor and adversary, revered and feared. But today, these communities face a new reality: nature refashioned as a commodity by Western capitalism. Rural subsistence economies are not exempt from the grinding pressures of population growth. Poverty, insecurity, displacement, and resource depletion are chronic risks, and daily survival is not assured. For those who still live "in" nature, such as Kekopey farmers and Maasai pastoralists, nature has become a means to an end and a path to progress, an attitude shared, ironically, by the policies of Western pro-development agencies. When survival is at stake, local people may have no choice but to tap natural resources. In Mukogodo, trees are burned for charcoal, sand is harvested for urban construction, and wild animals may be killed as nuisances or for profit. None of these practices are "good" for the natural environment, but they are often necessary. That means local people often view their land and "nature" as exploitable resources, an attitude echoing the Bible's problematic imprimatur for newly created humankind: "Let them have dominion."

I have seen firsthand in Kenya the very dynamics I'd read about while researching concepts of nature. I had witnessed population growth infiltrating baboon country. Inescapably, more people means more land taken for human use and the "humanization" of landscapes. Then follows human-wildlife conflict, which I first saw when baboons became crop raiders. It's an inevitable consequence of transforming wilderness into rangelands, farms, villages, and cities. Then there's the matter of wealthy developed nations imposing conservation policies on less advantaged parts of the world. Designed to promote biodiversity and protect natural resources, these policies predictably cause local communities to feel threatened and increasingly to push back, especially when new rules favor wildlife over human needs.

Attitudes about baboons have also shifted. Kenyan law in the 1970s changed baboon status from "wildlife" to "pests," and I understood why. Many Kenyans I have met, Black or white, consider baboons annoying at best, menaces at worst. When wildlife impedes human activities, it's bound to be disliked, then imperiled. I often ask local folks why they're so anti-baboon. They tell me it's not just that baboons take things but, worse, that they outsmart almost every attempt to keep them away. When they keep to themselves, baboons can be ignored, but when they venture into the human realm, conflict results and no one wins. Take "lodge baboons," who have lost their fear of people and expertly compete for tea tray goodies. Even I have to admit they can be nuisances. Lodge staff hate baboons, but tourists are charmed. Visitors are largely to blame for bad baboon behavior; they ignore the "Do Not Feed" signs and offer food scraps to the adorable monkeys. But the "pest" baboons always pay the price, never the humans who taught them to be pests. Baboons in the wild are normally different creatures, so when we view them as pests, we need to look at ourselves. Baboons can be cute, annoying, or even vicious, depending on your perspective. Again, it's complicated, as you'll see in the final chapter.

So, my research and my Kenyan experience showed me two basic and opposed worldviews. Sometimes they exist side by side in the same culture. One deems nature a sanctified and self-contained entity, the other a resource for human use. Here in rural Kenya, I walked a tightrope between them. Whose rights mattered most? I couldn't argue with the poor farmer or pastoralist who viewed wildlife as a competitor and threat to basic survival. And what right did I have to stop people harvesting sand or making charcoal to keep food on the table? Whose rights carried the most moral weight: those of baboons or people? Any judgment about the pros and cons of human impact—including crop raiding and the cactus invasion—will turn on your point of view.

---

I ask myself the same questions over and over again: Who speaks for those who cannot speak for themselves? Who had the right to the land at Kekopey, baboons or farmers? Baboons were there first, but the farmers hadn't meant to ruin their lives; they only wanted to be self-sufficient. As I delved more deeply, I saw a situation devoid of black-and-white clarity, with no good guys or bad guys. And I asked myself, What really is this thing we call "nature" in the Human Age, when humans, not nature, are the major evolutionary force?[6]

---

The more we learn about the animal mind (chapter 14), the harder we need to think about its implications. Does evidence of sentience change our relationship with animals? Does it demand a rethinking of nature? If animal mind is real, then animal suffering, previously discounted, might also be real. If so, aren't we obliged to ramp up animal welfare efforts? Moreover, we need to ask if mind endows animals with "personhood." If so, might they also deserve legislated rights? Questions like these inspired a respected international group of scientists and philosophers to form the Great Ape Project.[7] Peter Singer, a prominent bioethicist, leads the project and admits to using great apes as a wedge to open the conversation. As he had hoped, arguments about animal welfare and animal rights soon enveloped other species as well.[8]

The case for animal rights rests on two premises. In the first, philosophers like Singer and Bryan Norton employ the classic rationales for human rights to argue that rights should also belong to nonhumans. In his 1989 book *The Rights of Nature*, historian Roderick Nash traces the roots of this reasoning to liberal sensibilities first expressed in the Magna Carta.[9] Nash notes that rights claimed by baronial classes in 1215 were soon sought by less privileged classes. Thus began a trajectory that has strongly informed US history, from the Revolution in 1776 to the Civil War, up to more recent labor and civil rights movements. Almost invariably, public pressure has resulted in reforms and legislation.

Nature, too, was swept into this movement. With the creation of the National Park Service in 1916, nature won its first legal protections. Additional gains were codified in the 1970s with the Endangered Species Act, the Clean Water Act, and creation of the Environmental Protection Agency.[10] Nash observes that the granting of new rights both follows and reflects cultural shifts in attitudes, behavior, and morality. Ethics then come into the picture, providing a framework for morality. The progression is this: Changes in attitudes beget changes in behavior that lead to new definitions of morality, and finally to codifications in ethics that tell us what is and is not morally acceptable. His argument that nature, including animals, *should* attain rights (from humans), requires this progression, powered by social momentum. Nash imagines that ethics, as they evolve, will confer rights to *all* life: ecosystems, the inanimate natural environment, and eventually the planet.

There's a second premise behind the push for animal rights, this one rooted in the animal welfare movement and its focus on captive animals.[11] If we recognize advanced mental capacities in nonhuman primates and other captive animals, then indeed we must admit they feel pain and they suffer. New evidence that animals *do* suffer, coupled with an attitude shift toward empathy, has fueled reforms in the way we manage animals used as research subjects and food sources. Regulations now govern the care and maintenance of primates in US biomedical research facilities. In 2015, chimps were banned as research subjects in the United States, five years after the European Union banned *any* great ape from medical research. Primate "retirement homes" have sprung up, offering refuge for former ape and monkey research subjects. Concerns for the welfare of orphaned chimps, pet chimps, mistreated chimps, and other great apes spawned the Pan African Sanctuary Alliance movement, first in Africa and now elsewhere.* In the United States, the treatment of food stock animals has come under scrutiny by rights groups, resulting in greater regulatory oversight and somewhat more humane methods.

I thought about this chain of events, how findings about animal mind led to the current moral-ethical-legal reckoning. Just changing one characteristic of a species forced humans to rethink long-held and deeply embedded assumptions. But each halting step toward restorative justice for nature and wildlife comes with consequences, some predictable, some not. A regrettable truth is that the giving of rights to one group often happens at the expense of another. That means resistance is a part of a process that never

*For more on the Pan African Sanctuary Alliance, see https://pasa.org.

ends—whether you think nature has a prima facie right to exist, unmolested, or should be preserved for the benefit of humans, to enjoy or exploit. We think of conservation as purely good—and it is mostly good—but we ignore its complexities and consequences at our peril.

---

The media in all its forms has played a key role in reshaping attitudes about nature, and especially wild animals. We know that as attitudes change, behavior changes. Documentaries and images in popular publications have made chimps, elephants, and countless other animals—even baboons—part of our moral universe. Starting in the 1960s, Jane Goodall documentaries were a surprising bonanza for National Geographic, one that continues to this day. The intent was noble: to open a secret door to the natural world and invite viewers to marvel at its wonders from the comfort of their living rooms. Animals were the special draw. People identified, fell in love, and rallied behind animal rights and conservation.

Despite all the good they did, these great documentaries had a pernicious side effect. As feelings in the United States and Europe about wildlife and nature grew warmer, sympathy for nature-adjacent people cooled. Yes, the animals may have been threatened, but all too often so too were the local people who shared their habitat. When human welfare impinges on animal lives, the international media firmly sides with the animals. Here's something few people know. In Kenya today, more people are killed by elephants than elephants are poached by people. The media and some conservation organizations don't like this fact and won't report it. In a dismal irony, as the media humanized wild animals they often dehumanized, even demonized, local people.

Human-wildlife conflict, the thorniest of current conservation crises, rarely made it into "blue chip" nature documentaries until recently. Indeed, the ease with which a media outlet can manipulate a narrative and bias perceptions was something I learned, to my regret, firsthand. In 1996, a company (dubbed Perverse Productions by me) set about filming women who study wild primates. By this time, the baboons and I had appeared in 15 respectable documentaries; I had little reason to think this time would be different. The film, we were told, would showcase and celebrate our scientific achievements. The finished product, however, was jaw-droppingly different. Now titled *Beauty and the Beast*, the documentary's footage had been surreptitiously cut by London-based editors to portray female primatologists as women driven by raw maternal instinct. Our subjects were hairy substitutes for the human babies we longed for. It was then I realized that the story line, the narrative—*not* the images—is what shapes a viewer's response. Media companies had become self-appointed gatekeepers of popular ideas about nature, wildlife, and conservation.

Another example of selective editing proved equally instructive. I had heard about two versions, American and British, of a documentary titled *Earth*. I decided to show both to students in my conservation course. Visually the films were identical, but the narratives were wildly different. The British version was serious and respectful of its audience's presumed intelligence. The American version was a treacly insult. The wildlife "stars" were cast as 1950s sitcom stereotypes, drama was hyped, and emotions were pandered to, up to the pulpy happy ending. It was hard to know what was most offensive: the presumption of American stupidity or the distortion of an important message. But here was confirmation of two personal convictions: that cultural differences are real, even when the cultures are historically close, and that media narratives powerfully mold perspectives, for better and sometimes worse.

---

It's easy to complain about all that's wrong in the world, but it's not very helpful. Action is needed, and that includes changing how we think about imperiled "nature." What exactly are we trying to save? I think of the well-intentioned scientists who proposed a "rewilding" experiment. The plan was to restore a portion of the pre-human North American landscape and its wealth of wildlife* as a counterweight to typically doom-laden scientific news.[12] But were they realistic? Not only did they neglect to consult people living in the proposed project area, but they failed to take into account vast historic transformations in the habitat. The soil, air, water, climate, vegetation, and wildlife were all very changed since humans first made their way to the North American continent some 30 millennia ago. There simply is no going back.

Jonah, once again, offered me a different perspective. His book *We Alone* redefines conservation, tracing its origins from the earliest gatherer-hunters through all of human evolution.[13] Looking for reasons to hope, he finds it in the new idea that conservation is about more than preserving biodiversity or "saving nature." Humanity needs help too, and it *can* help. He challenged me to consider a world where chimpanzees or baboons or elephants—not humans—were superdominant. These species wouldn't care about each other's fates, or that of "nature," or of the planet. This approach struck me as no less than revolutionary. Humans created the mess we have, but Jonah argues that we need to stop seeing humans as inherently destructive. Humans are trying to fix things—at least some are—something no other species has ever done. Jonah's reasoning echoes a quote from the great evolutionary thinkers J. B. S. Haldane and Julian Huxley: "The one great difference between man and all other animals is that for them evolution must always be a blind force, of which they are quite unconscious; whereas man has, in some measure at least, the

*Their calculations regarding wildlife and the ecology were based on the fossil record.

possibility of consciously controlling evolution according to his wishes."[14]

Finally, Jonah convinced me I was asking the wrong question when I wondered about what things in nature most needed saving. Instead, he urged me to focus on ways that people and nature can meet in the middle and learn to coexist. I liked the idea of discarding the win/lose binary and recasting it as a win-win, even when painful compromises were necessary. But for that to happen, broadly and sustainably, humans as a group may need to scale up their "humanity."

In graduate school I made the decision to study baboons, not people. At first, the baboons insulated me from people. Soon, however, I realized that the future of the baboons I cared about was tightly bound to the future of their human neighbors. As long as the Maasai were marginalized and poor, the baboons were at risk. Clearly, we needed new and better solutions to old and new problems, because none of us could go back to the old days, not the baboons, not the Maasai, and not me.

## Chapter 20

# Coexistence in a World in Flux

**I WOULD LIKE TO PROPOSE** a new way of managing our relationship with wildlife, one that rejects the winner/loser paradigm we take for granted. Though I never embraced this model, for a long while I accepted it. After decades in the field with baboons, and with guidance from Jonah, I realized there was a better way. It is *coexistence*, a concept that feels obvious and commonsensical but that rarely gets implemented by mainstream conservationists. I believe it's the best hope we have for sustainably balancing the needs of humans and wildlife. But it's no easy fix. What does coexistence *really* look like? It means people collaborating, compromising, and, yes, using science, to guarantee the equitable sharing of the world and its resources with animals. My thoughts began to evolve early on, when I realized that hating farmers for killing baboons was counterproductive. You may recall the Kekopey farmer who invited me to "tea" to discuss his predicament, offering his last maize cob in a gesture that nearly made me weep. Baboons had taken many of his cobs, but he didn't hate me. Then, to offset the income lost to the baboons, I began helping farmers with the Woolcraft weaving project. Which is to say, in this world in flux, coexistence takes real work. Even the well-intentioned may resist its necessary compromises and concessions. I'll start with a cautionary tale, but later I offer hope.

In 2012, I got a call from Justin O'Riain, then head of the University of Cape Town's Baboon Research Unit. He invited me to present my crop raiding findings to a group of concerned scientists. They had been tasked with creating solutions for an out-of-control baboon situation in this South African city of four million. I had heard rumblings about Cape Town's urbanized baboons but hadn't realized things were so bad. The city had become home and smorgasbord to several troops then totaling more than 400 chacma baboons, whose males can weigh up to 70 pounds. I knew from the translocation how baboons readily adapt, but the animals I had moved were adjusting to harsher conditions than they'd known. In Cape Town, the situation was quite the opposite. Baboons had settled into a veritable wonderland of resources. Their natural habitat, botanically rich heathlands called the fynbos, had been claimed by urban and suburban development. Displaced, these savvy baboons had learned to check for unlocked doors and open windows—both in cars and homes—where human food often awaited them. They rifled through garbage cans, feasted on the edibles, and left a mess. Rooftops took the place of sleeping rocks, and baboons were pooping everywhere. Baboons were even lying in wait to get into homes. In one widely reported case, a male baboon ambushed a pregnant woman as she unlocked her front door after grocery shopping. Groceries went flying as the woman dashed for safety to her bedroom. The baboon, meanwhile, helped himself to both groceries and fridge.

But nothing quite prepared me for seeing the situation firsthand. Cape Town had baboon protocols, but they weren't working. I had never witnessed such atrocious baboon behavior, but, as I saw it, the animals were not to blame. Justin planned a full agenda to get me up to speed. First came a workshop with fellow scientists. There I learned that the "experts" working on the problem were ignorant of community-based conservation and hadn't met with resident stakeholders. Indigenous communities hadn't lived in South African wildlife areas for a century, perhaps the reason community work hadn't become part of their cultural DNA. In one outlying private park, Tswalu, conservation efforts were ongoing. The work had successfully restored some missing biodiversity, but the link between people and conservation had been lost. No wonder South Africans and Kenyans never see eye to eye at international conservation meetings. Kenyan Maasai have a voice and a say in rangeland management, while Black Africans in South Africa have neither. It matters, because community involvement is the linchpin of coexistence. Cape Town itself was extremely diffuse, a "community" only in the broadest sense, more a collection of self-contained neighborhoods with varying levels of baboon activity.

The next meeting was with a "community" of sorts—"white" South Africans, clearly under the sway of white animal welfare activists and their media blitz. These activists opposed on humane grounds all the standard deterrence methods—loud noises, paintball guns, and the culling of incorrigible males. Indeed, the culling of male baboons had sparked their activism. At their insistence, a "gentle" approach was now in use. Paid baboon monitors were only allowed to clap as they shooed baboons away from human food sources. I got their point that baboons had a right to territory taken from them by people, but it seemed a poor excuse for enabling bad baboon behavior. I offered the key lesson from crop raiding: For baboons to be deterred, the cost of obtaining human foods had to exceed its benefits. Clapping monitors did not represent a serious cost, barely a slap on the wrist. The predictable outcome was increasingly emboldened baboons. I made my points, took questions, and believed I had convinced some to view the problem from a new angle.

But the true eye-opener was the next day's tour of the "suburbs." There I saw monitors clapping their hands in vain attempts to keep baboons from raiding "baboon-proof" rubbish cans that people had failed to secure. We watched a troop descend from an apartment roof where they had slept next to heat vents, scattering people in the halls as they lit out. Then we visited a bakery, another sleeping spot, and saw baboons feasting on handouts. Clearly, these were established ways and would be hard to change.

At town hall meetings, it struck me that the audience was "white" and mostly affluent—both pro-baboon activists and those who wanted baboons banished, or worse. Though very unlike the poor farmers and pastoralists I was used to, the anti-baboon contingent had similar complaints. They'd invested in a garden or a house—or in one case, a commercial vineyard—and baboons were causing damage and expense. What did I suggest? I pushed back on the activists, insisting that harsher measures were needed. Individual animals might need to be sacrificed so the group would learn. Seeing one of their own killed while raiding would deter the others. Could Cape Town baboons be controlled and peaceful coexistence achieved? I didn't know. The odds looked slim because urban raiding was a fixed tradition now, passed from one generation to the next.

Our last day featured a field trip to outlying areas frequented by tourists and also problem baboons. We stopped at a wayside observation point and, sure enough, as soon as a group of tourists gathered, baboons materialized. At first, the tourists were excited to see wild baboons up close. A young German mother stood near her open car door holding a baby and I knew what was coming. I called to her to get into the car and lock the door, or at least to lock the door if she was going to stay outside. But she stood her ground until it was too late. A male baboon climbed into the car and went straight for her backpack. He carried the pack outside, searched the contents, and settled down to enjoy her lunch. Only now did she scream at him, but he barely noticed. His meal finished, he methodically proceeded to the next car, and the one after that. Unfortunately, I learned later that this bold male was marked for culling.

The problem was, as ever, with humans who subverted the "nice" baboon behaviors I knew by failing to control baboon access to human food—through ignorance, indifference, or misguided caring. Every conflict situation I knew had roots in human activity. Well-intentioned activists were responsible for turning years of bad baboon behavior into a tradition. Even if the natural world suddenly offered all the food they needed, the baboons would persist with urban raiding. Entrenched behaviors are as hard to change in baboons as in people. Not helping the matter was an urban raiding cost-benefit ratio skewing massively toward benefit. Baboons helped themselves not only to picnic lunches, groceries, and garbage bins, but fruit trees, kitchen gardens, and free bakery goods. This was coexistence run amok.

I left Cape Town with a heavy heart. It seemed a dismal preview of a dystopian future where people and "wild" animals coexist in disharmony and imbalance. In this case, it was a zero-sum game that baboons were winning at the expense of people, provoking an inevitable and dangerous backlash. Only time would tell if my advice would be taken and help.

---

Emerging from the urban cauldron, I had reason to reflect on our successes in this rural corner of Kenya. I was returning to our thriving project at Twala, where a group of wonderful Maasai women—a true community that taught me much—were nurturing a set of small businesses modeled on community-based conservation. The result was coexistence. Granted, we did not face the challenges of a sprawling metropolis, but Twala is a place where baboon raiding is more than a nuisance. It can bring people closer to hunger.

My work at Twala began in 1996, when eight Maasai women shared with me their idea of a cultural *manyatta*, or homestead.

I had been invited to their meeting by Joice Mamai, whose face, like those of many Maasai women, reflected a hard life and a buoyant spirit. With her eight children now in school, she alone fetched fuelwood and water, cooked and cleaned, and tended the family's herd of sheep and goats. All the women were mothers and busy multitaskers, yet they'd managed to hatch an ingenious plan. The cultural manyatta they had in mind would preserve Maasai traditions and ideally earn some money. They wanted my support. While I loved the idea and their enthusiasm, I had doubts. Twala—a settlement named for a nearby dry river—was not on any tourist route, but only tourists could provide a revenue lifeline.

Thus began a rich and rewarding collaboration. Benefits flowed not only to these brave and enterprising women, but to the wider community and to local baboons as well. But the first order of business was to figure out how to attract (well-heeled, we hoped) visitors. I floated the idea of *eco-walks*: walks with baboons and livestock, and tours of local plant life. I would help with the baboon feature while their deep knowledge of livestock and plants would be put to good use. The eco-walk idea went over well and we had a plan. Now to put it into play.

For the baboon walk, we needed to habituate a local wild baboon troop to the unpredictable presence of tourists. We selected a medium-sized troop, Sisal, who were the bane of Namu's life previously. Baboons naturally run from unfamiliar humans and require slow, incremental desensitization. Once you have gained baboon trust, you are a "habituator" and what we call a "passport," whose presence assures baboons that a stranger means no harm. Rosemary—sharp, bilingual and soon to be a great asset to the project—volunteered as my first guide trainee.

A sort of magic happens when you're with baboons, and it's been felt by virtually every person who's joined me in the field. Imagine if you will standing unnoticed in the middle of 30 or 40 baboons as they swirl and tumble around you, playing or showing off or testing their standing in the group. Maybe they're just digging for roots, munching blossoms, or blissfully grooming each other. It's like a reality TV show, both comic and dramatic, only better. Rosemary was instantly smitten. A family drama unfolded as an avoidant mom led her distressed infant, Doc, on a chase around a tree. I provided a running commentary. We cheered when the mother's male friend finally intervened and settled everyone down. With mother calmed, Doc climbed onto her belly and latched on for a feeding. Entranced, Rosemary declared that she just wanted to watch baboons all the time. I wished she could, but now she was also the hospitality manager for Twala. Could it be that the local community was an untapped source of baboon support? Perhaps to know baboons was to love them, and that gave me hope for harmonious coexistence.

The livestock walk relied on knowledge of Maasai herding practices. I was surprised to learn that younger Maasai men knew little of those ways. Pastoralism seemed to be slipping away. So we recruited older men for the livestock walk, even though they (wrongly) doubted that their traditional knowledge would be fascinating to "outsiders." Women would lead the plant walks, focusing on medicinal plants used by Maasai to treat all manner of ailments. Each went on to generate much needed revenue, but Rosemary's baboon walk proved the biggest hit. So far, so good.

I was definitely outside my comfort zone. I am a scientist, not an entrepreneur! But the trial-and-error approach to moneymaking endeavors was not wholly unlike hypothesis testing. Following the axiom "When life gives you lemons, make lemonade," we experimented with juice from the *Opuntia* fruit, the same cactus fruit relished by baboons. The plan was to sell the ingredients for *Opuntia* juice "cocktails" to high-end tourist lodges around Kenya. But first we had to create the cocktails. Maasai women abstain, so it fell to Carissa (now grown) and me to concoct a few recipes. (Someone had to do it . . .) We finally settled on two: Drunken Monkey—a mixture of vodka, tonic, *Opuntia* juice, and ice—and Twala Sunset, a play on tequila sunrise featuring rum, passion juice, *Opuntia* juice, and soda water. A popular lodge in Amboseli National Park agreed to host a test run. In preparation, we posted signs inviting guests to enjoy a cocktail while supporting the efforts of local women. The cocktails were a hit but, sadly, short-lived. A severe drought set in and cactus stopped fruiting. Meanwhile, Maasai women had begun taking advantage of the juice, adding it to their children's porridge and even making tea from it. *Opuntia stricta* fruit, we learned, is high in vitamins, minerals, and antioxidants, and even helps control diabetes. This small addition had greatly improved their diet, which lacked fresh fruits and vegetables. But for now, we all had to wait out the drought.

By 2010, Twala was turning a profit. Success attracted donors—including the Dutch government, the European Union, and World Vision—enabling the construction of eco-cottages, a meeting hall, a kitchen, and a curio shop. Guests tried their hands at herding livestock the Maasai way, and mostly failed, but everyone had a good laugh. On the plant walk, visitors marveled at both plant diversity and the Maasai women's deep knowledge of their medicinal uses. But those who dared nibble a plant were always shocked by its bitterness. The women gradually added cultural activities: jewelry beading, cleaning milk gourds, bee keeping for honey, and harvesting aloe for soap and cosmetics. Men pitched in as well, showing how to make bows and arrows and teaching a board game called *bao*. In the most popular cultural activity, visitors attempted a typical Maasai woman's daily chores, including fetching water. Large men visitors struggled with the standard five-gallon container, full to the brim, and few women visitors even tried. Then a diminutive Maasai woman would step up and heft the container as if it were a tiny bucket. As expected in Maasai culture, she'd be modest about the feat, but we were all proud. It was a good reminder of what we from the developed world take for granted, such as indoor plumbing.

It was and is essential that I, as a scientist, remain politically neutral. But I couldn't ignore the impact that Twala activities were having on the women. Some flouted a cultural taboo and began speaking in the presence of men. Many were earning their first money and getting a taste of financial independence. A few summoned the courage to leave abusive husbands. It was a case study in female empowerment, and it was happening, at least indirectly, thanks to baboons—now considered assets, not liabilities.

I once eavesdropped as Rosemary showed guests around. "We want to preserve, not destroy, Maasai traditions" she said, "but we also see that many traditions harm the girl-child. We don't want those to continue." This was a brave statement coming from a woman who belonged to a strongly patriarchal society that still practiced female genital mutilation, condoned child marriage, and opposed education for girls. Lucky girls with enlightened fathers found themselves in schools still dominated by boys. Those girls began to imagine a different future for themselves.

Twala flourished because of word of mouth and brochures distributed by the indefatigable Rosemary, who'd adopted the unofficial role of Twala's ambassador. We also secured a helpful place on the African Conservation Centre's website. I love to watch the song and dance performance that greets Twala's new guests—a delightful and endearing reception. One day, the welcome song seemed to be about me. Curious (and slightly mortified), I asked one of the women to translate. My heart melted as I heard words of praise and gratitude for my help. The song also solved a chronic dilemma. Up to now, the women hadn't quite known what to call me. Sometimes it was Strum, sometimes Dr. Shirley, other times Shirley became Charlie or something else altogether. But in this song the woman had christened me *Mama Twala*, the mother of

Twala, which meant I was no longer an outsider—*mama nugu*, mother of baboons—but one of them. Now I have the honor of saying I am Mama Twala, which is all that's needed.

---

Any talk of coexistence must take into account the ecological and social pressures on both humans and wildlife.[1] Unfortunately, the Baboon Project's ecological monitoring showed the same erosion and degradation patterns that Twala women had noticed outside Twala's 40 acres. Our documentation began in 1984, when the baboons and I first moved to Laikipia. Over time, we had captured striking changes in the ecosystem, from Il Polei Centre to the little village of Doldol. When the study began, the Maasai had ample space for migratory pastoralism. People and herds arrived during the rains when grass was abundant and left once the pasture dried. Coexistence with baboons and other wildlife was seasonal and relatively easy. But the El Niño rains of 1998 changed all that. Grass grew so lushly that cattle no longer needed herding to new grasslands. There were other benefits. Once settled, people could avail themselves of a new free medical clinic, a great luxury in this marginal place. Improvements in the local primary school tempted some families to educate their sons. In normal years with less rainfall, a boy's education would break off when his family left the area for new pasturelands. Graduation from primary school might be at age 18, if at all. Other forces began to spell doom to migratory pastoralism. National parks and private owners were fencing more land. Both took dry season pastureland out of play. Wildlife was also losing habitat, but herders and livestock lost more.

As temporary homesteads became permanent, people and their animals were squeezed into smaller areas, degrading the quality of rangeland. I watched as livestock heavily grazed these smaller pastures, and I recorded the downward spiral of grass production. As rangelands shrank, herds grew smaller and less able to sustain families. People resorted to sand harvesting and charcoal making just to get by. Game trails were becoming erosion gullies as bare soil washed away, leaving an unstable pebbled surface. Baboon tracking became more physically treacherous for me in this altered terrain. Plant and wildlife diversity declined as soil quality worsened. Stands of large euphorbias and the big acacia trees disappeared entirely. We seldom saw large predators, a mixed blessing, and the wide variety of birds, once ubiquitous on communal land, gradually dwindled to only a few species. These new and unwelcome conditions challenged every life-form on the savanna.

The people of Mukogodo were understandably distressed by these transformations and not always receptive to science-based explanations. So I wasn't terribly surprised when baboons began to be blamed. Sometimes baboons do cause harm, but, as always, problems only happen when people choose to settle in baboon terrain. Now it seemed that a single baboon transgression could erase all of the Baboon Project's benefits to the community. If a male baboon took a goat, I was responsible. Didn't I own the baboons? No, I didn't. In Kenya, wildlife belongs to the people under the custodian ship of the Kenyan Wildlife Service. But I understood why people might have thought so. Baboon predation on livestock is rare, but indiscreet. I dreaded the times a baboon took a goat and I could do nothing about it. But I had learned an important lesson on Kekopey: Respectful listening always turned down the heat.

During the crop-raiding days on Kekopey, all possible outcomes seemed to have winners and losers. But Jonah's insights about the longtime coexistence of Maasai and wildlife put me on a different track. Here in Mukogodo we needed to employ the same conciliatory approach and make work in the community a top priority. The Baboon Project became a driving force behind partnerships for development and conservation at Twala and with the schools. More children were getting educated and envisioning lives beyond pastoralism. And, slowly but surely, people were starting to like baboons. A wildlife attitude survey conducted by the African Conservation Centre, taken across Maasai lands in Kenya, revealed positive attitudes about baboons among Maasai in the Mukogodo community where I worked. Elsewhere, including areas where other baboon studies were ongoing, anti-baboon sentiment prevailed. Coexistence requires humans to share the world with other creatures, even if it entails sacrifice. It's an approach that cultivates compassion and, importantly, opportunities, so both humans and wildlife can flourish.

---

How might coexistence work in an urban setting like Cape Town? Before my visit, I believed coexistence was the only approach to wildlife management. But is coexistence even plausible in such a place? The special challenges in Cape Town were obvious. Not only is it geographically sprawling, but its vast populace is diverse in every sense. Unanimity, or even consensus, about how to solve a problem like urbanized baboons is unlikely. As with all conflict between wildlife and humans, the problem lies with the human presence and human behavior. In Cape Town, people weren't making consistent, proper use of baboon-proof bins or equipping baboon monitors with deterrence methods that worked. Electric fences were an option only for wealthier communities. I wondered if they'd ever make a unified effort to keep baboons away from human food, the only path to success. In a city of millions, maybe not, especially when tourists named baboons as one of the favorite features of the city. The problem is this: Baboons are too smart not to make the most of the best food resources available, especially when those resources are nutritionally dense human food.

In the aftermath of my visit, I sent a blunt letter to the Cape Town newspaper. In it, I blamed the activists for indulging the baboons' worst tendencies. In a grim irony, they'd assured the culling and killing of the boldest baboons, but the problem persisted. People took heed. Monitors were allowed to use noisemakers and paintball guns, and it helped. According to recent reports, things are better. The Cape Town baboon population now numbers around 600 individuals in several troops, up from 400 ten years ago. Some baboons have left neighborhoods they once terrorized, and authorities have adopted a baboon birth control program. I think of when Carissa was little and a python took one of our beloved domestic cats. We were both upset. Then, through tears, she turned to me and said, "After all, the python was here first." Coexistence means letting go of expectations that humans can do whatever they want without constraint or costs. I wonder if there will be a time when most people on the planet see the world as my wise little daughter did.

---

I run my own little coexistence experiment living on the edge of Nairobi National Park. Jonah and I moved into a "wildlife" area, which was much of its appeal. Here we would have our own wildlife bubble on the edge of Nairobi's big urban space, buffered by Nairobi National Park. Wildlife was here first, so what would coexistence look like? First, we had to fence our property, not to keep out wildlife, but to secure wildlife from humans and livestock. Then we needed to modify the area close to our house. Wildlife was thriving across the property, but a few animals required special tactics. For instance, hyraxes, rodent-like creatures whose closest living relative is the elephant, were eating everything we planted around the house. So we relandscaped our yard with plants that hyraxes had chosen not to eat—only now they seem to be developing a taste even for those. Later we planted a kitchen garden that instantly attracted baboons, vervets, and birds, a problem solved by a chicken wire enclosure. Now we enjoy lettuce, tomatillos, and black raspberries while wildlife has free rein to forage elsewhere on our land.

Coexistence with wildlife involves other compromises. I don't walk alone when buffalo are about. They present a special challenge because my fused back requires me to watch carefully where I step. If I were to surprise a sleeping buffalo, it would likely charge; I certainly can't outrun a buffalo. It also means letting Jonah "remind" the baboons and vervets that they belong in the National Park and not on our side of the gorge. Sometimes he uses an air rifle to pepper the shiny, large bottoms of male baboons. I don't watch, but I tell myself it doesn't really hurt them, only stings. Vervets get a different treatment, being too small for air rifle pellets. Instead, Jonah tosses small stones at them (not hitting them) until they retreat into "their" space. We take consistent measures to prevent unwanted wildlife incursions from becoming "tradition," mindful not to cause harm. It's a worthwhile effort that has become routine. Since native animals were here long before we arrived, the work of compromise falls on us. That's only fair, and the payoff is great. Coexistence with these creatures is endlessly fascinating and less bother than the daily work of caring for our pets.

---

As an educated white woman with a pension from a progressive university, living comfortably in the modern global village, I have no need to grow food or herd livestock to survive. Large, dangerous animals still inhabit my world, but I can find safety in my house or car and enjoy their presence without fear. *Clearly, my privileged position makes coexistence easier*. Yet, in this human-dominated world, my privileged point of view underscores what is needed to make coexistence possible. Westerners, as they became wealthier, healthier, better educated, and more urbanized, began to view nature less as a resource to exploit and more as a system with a *right* to exist (chapter 19). The women of Twala are changing as they embrace new opportunities and resources that bring them better education, health care, employment, and security. The baboons play a role in this scenario, being central to the women's livelihoods and growing independence. The point being, as the quality of these women's lives improves, the baboons' future looks brighter, in an ideal scenario of coexistence.

Coexistence gives me hope when I think about the past and the possible future for baboons and other wildlife. Baboons and humans have much in common, which contributes to conflict when they meet. Baboons are smart generalists with primate hands, big brains, and complex social groups. Humans are even smarter generalists with even bigger brains who took their primate hands and brains and added tools and culture to become not just a successful primate, but the predominant large mammal on earth. Cooperation among unrelated individuals along with tools, language, culture, and planning liberated humans from the evolutionary constraints that apply to other species, those of body, kinship, and the environment.[2] New social emotions developed—feelings like love, pride, gratitude, and shame—and added dimensions to our novel society. Humans took nature apart, separating things that had been bound together for eons, taking oil, minerals, and trees from the earth and using them to reengineer the natural environment. As our tools became "technology," we reshaped the world with our cities, dams, harbors, roads, rails, and sprawling agricultural tracts.

What will the future bring in this new Human Age, the Anthropocene?[3] Our human footprint is everywhere, and everything is more connected than ever before. This means that to save the baboons I study, I need to imagine them at the center of a vast web of interlocking forces that may one day impinge on their landscape and resources. Some already have. Above all, I must consider their human neighbors, because conservation success in the

Anthropocene depends on caring about the future of people as much as caring for the future of wild species. At the end of the day, people hold the key to the survival of all living creatures.

The baboons I first watched are gone: Peggy, who taught me the importance of family and hierarchy; Ray, who showed me why even powerful males need friends; Wiggle, who defended his friends to the end; and Zilla, that wise old female who never gave up. These are the animals I knew and still miss. But also gone is the world as they knew it. The question is whether we humans will do what we know is needed to protect nature and wildlife and make, when necessary, unwanted sacrifices. I'm optimistic because we are *Homo prospectus*, a species that can think about the future.[4] I have already seen a change in the university students I teach. They care about both the planet and the people. They focus on solutions because they already know the problems. For now, flexible, adaptable, smart baboons have made their way in our humanized world, with mixed but always interesting results. How long will they remain one of Africa's most successful wild mammals? The answer depends on how we humans see baboons in our future.

Appendix 1

# Sketches of Baboon Eyebrows

Kind

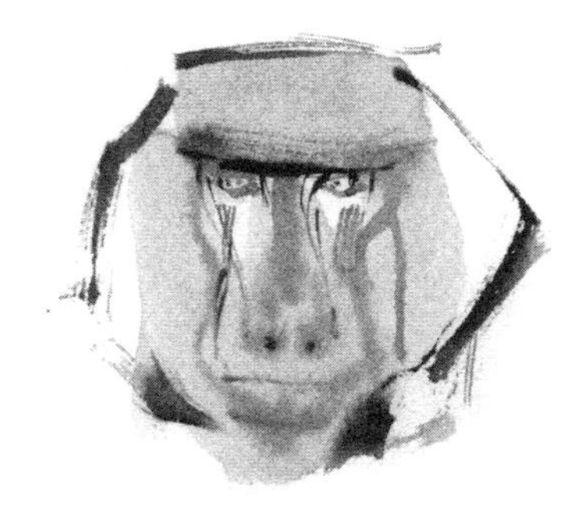

Frightening

Appendix 2

# Data Sheets

670

A 1
D
8
F 11

5/7/79 PHG PA/PT/R/NT

7:00 cool cloudy still
P37 on cliffs - some wh
7:10 descend cliffs for gr
and mushrooms

[How long before BE w/ from again after infants born?]

7:15 HO cop TH −/+ +/+ $S_3^+$
HO gr TH
7:20 GA ~ FA from MQ
7:40 HO cop TH $S_3^+$
MQ GA +
8:15 HO cop TH $S_3^+$
MQ RD + cloudy
MQ > HO as HO + TH
is this a reversal ?
9:04 SP tries to enlist O against www
SP gg thr e www
www ≈ SP
SP sn gg www
SP → gr DA as
DM gr DA
www > DA
SP/www DA/www
9:40 GA walks from other side O
- agg voc - TURNOVER
GA cop TH $S_3^+$
10:10 GA cop TH −/+ +/−
HO 72+ GA gg ~ HO
10:35 BE gr W x
QT → www
NE → www
QT thr e BE ø
NE thr e BE ø
QT atag BE
BE sn > QT
BE thr e ≈ QT
QT > BE
QT → enlist SB
QT SB thr BE øBE
BE ~ QT SB
QT SB > BE
BE return to W
W gr BE
pi QT = BE

10:38 LY → ~ BE from www
QT + LY
QT protect the BE
BE >gg LY QT
LY → www
QT gr LY
www ~ LY
KIN/pi LY/BE QT/BE

10:50 P40 ~~all to~~ harem
tommies n > 50
no young babies
vis AG → 25 + 50
T > OBS
11:00 ✓ RO + DV P6
11:30 GA ~ TH &
got mushrooms
TH > GA
but keeps much.
11:40 GA cop TH +/+ +/+
12:16 IA cop WO −/+ +/− $S_4^+$
12:22 MQ BO BS agg HO
BO sn ≈ HO
HO > BO
HO → thr gg BO
LY gr www
BO sn P > HO
HO = BO
12:25 NT cop WO −/+ +/−

| Øgr | Øgr | SS | gr |
|---|---|---|---|
| NT/WO | WO/NT | HO gr TH | FA gr MQ |
| | | DA gr WI | SB gr SESS |
| | | DA gr CJ | SO gr SP |
| | | AD gr CL | FA gr V x |
| | | WO gr NT | MD gr M |
| | | GA gr TH | M gr MG x |
| | | AL gr DI (after agg) | DM gr MR |
| | | | VI gr V x |
| | | | TH gr TD x |
| | | | RD gr ME x |
| | | | ME gr MQ |
| | | | TD gr RD |
| | | | DA gr www |
| | | | DA gr DM x |
| | | | SI gr DA |
| | | | SM gr KI |
| | | | TH gr TE x |
| | | | KT gr KI |
| | | | SH gr COX |
| | | | TD gr JE |
| | | | BE gr BT x |
| | | | MV gr www |
| | | | JE gr JH x |
| | | | AN gr AU |
| | | | ZA gr NE |
| | | | CO gr BU |
| | | | AN gr AD |
| | | | SH gr CN |
| | | | Z gr ZA |
| | | | BE gr W x |
| | | | JE gr www x |
| | | | Z gr ZA x |
| | | | CY gr CO |
| | | | QT gr www |
| | | | QT gr LY |
| | | | ZA gr www ✓ |

FA >gg GA
JE >f TO
MQ >S HO
ME >S PA
PA >S RD
MD >S K
MD >gg RZ
M >S RZ
www >gg SP
SESS >f RZ
SP >S CL
DM >S CL
DA >S CL
RZ >f SESS
KI >S NT
RO >S LY
SESS >f HO
SH >S GA
CO >S GA
MV >S GA
NE >gg AD
AD >S Z
HO >gg GA
AD >S TH
BE >gg JE
www >S JE
www >S LY
BE >gg LY
WO >f DV
RO >S WO
DT >agg AL
WO >S DT BE
DN >S SHCM
SHCM >f TH
TH >agg GA
BO >agg HO ✓

# Adlib Codes

**Avoid Column:**

>— avoid
>—< mutual avoid
**s** spatial displacement
(one moves away from another's approach)
**fd** feeding displacement
(the displacer investigates or takes the food)
**agg**$^{1}$ threat (eye flash, slap ground, lunge)
**agg**$^{2}$ threat + chase
**agg**$^{3}$ contact aggression
**c** feeding displacement (involving corm sites)
**ff** fearful avoidance (fearful, geck, tail up, scream)
**fg** avoid a greeting (showing signs of FEAR)
**g**$^{1}$ avoid a greeting (NO contact)
**g**$^{2}$ avoid a greeting (WITH contact)
*Underline the individual who initiates the greeting.
**sg** social displacement
(groom partner, proximity to infant)
>— avoid sex interaction (use sex codes)
**sp** avoid rough play
**hd**$^{1}$ avoid herding (female avoid push from male)
**hd**$^{2}$ avoid herding (female avoids chase or contact aggression from a male)

**Groom Column:**

A **gr** B (A grooms B)
**X** Mutual Groom (grooming trade-off)
Four Possibilities per grooming pair:

1. A **gr** B
2. B **gr** A

3. A **gr** B**x**
4. B **gr** A**x**

Include infant with mother

**Other Columns:**

**SSgr** grooming including a sexually receptive female (with males of all age classes—NOT FAMILY)
**Ø agg** no response to aggression
A/B (A = screamer/B = aggressor)
♂**g** male greetings (Underline/Overline ID who initiates/leaves)

**Reconciliatory Codes (after aggression):**

**rc1** conciliatory—positive behavior (greet, groom, play, etc.)
**rc2** nonconciliatory—aggression, displacement

**Sexual Behavior Codes:**

**pr** present
**mt** mount
**½mt** half mount
**sn** sniff
**hg** hip grasp
**ps** push to stand as part of repro

**Other Codes:**

**+** follow
**U** walk around
↓ sit
∩ weaning
**!!** infant distress
**em** embrace
∞ take ventral
**dor** carry dorsal

| | |
|---|---|
| **la** | look at |
| **lb** | look back |
| **lw** | look away |
| **ht** | touch (hand to) |
| **gr** | groom |
| **g$^1$** | greet (no contact) |
| **g$^2$** | greet (with contact) |
| **g$^{2r}$** | rough greeting (to infant) |

*Overline the individual who leaves the greeting.

| | |
|---|---|
| **lp** | lipsmack |
| **Ø** | no response to (no reaction) |
| **ati** | attention to infant |
| → | approach |
| ←→ | mutual approach |
| **o/s** | out of sight |
| θ | troop |
| **wh** | wahoo (lost or danger) |
| **gt** | grunt |
| **dd** | diddle |
| **va** | veer away |
| **gt** | grunt |
| **L** | leave |

**Fear & Aggression Codes:**

| | |
|---|---|
| **gk** | geck |
| **scr** | scream |
| **mg** | molar grind |
| **pg** | pant grunt (threat) |
| **ya** | yawn as threat |
| **thr** | threat (slap ground, lunge) |
| **thr$_e$** | eyelid threat/flash |
| **ff** | fear face |
| **bb** | back bite |
| ↑ | tail up |

| | |
|---|---|
| ~ | chase |
| ≁ | chase not trying to catch |
| **nls** | enlist |
| **sup** | support |
| **nct** | incite |
| **rd** | redirect (aggression) |
| **hd1** | herding (push) |
| **hd2** | aggressive herding (chase) |

**Avoid Codes for External Interference:**

| | |
|---|---|
| >+ | no avoid |
| **1** | displacement |
| **2** | avoid |
| **3** | avoid strongly |

**NB: Write number over avoid symbol.**

# ACTIVITY SCAN

DATE: 11/17/23 (B - 1 -Activity Scan) TROOP: NMU WQs pg: 1/5

OBS: BJM

put as many scans on one sheet as possible

| TIME | ADULTS ♂ | ADULTS ♀ | SUB ADULTS ♂ | SUB ADULTS ♀ | JUVENILES ♂ | JUVENILES ♀ | INFANTS ♂ | INFANTS ♀ | LOC/troops N= /HAB |
|---|---|---|---|---|---|---|---|---|---|
| 12:30 | Wf. E2G<br>OI. E202L<br>Gv. 2E2G<br>fG. E202L<br>Hs. E2G<br>DG. E92G<br>FP. B (E02) | BQ. E92G<br>KT. E202L<br>CF. E92G<br>DU-ME202L<br>TW. E202L<br>DO. ME16L<br>HA. B (PQ)<br>CR. E14L<br>LQ. W | ZE. E92G<br>KQ. E202L<br>AM. E202L<br>KS. E202L<br>B2. E2G<br>EX. T<br>PG. E2G<br>AN2. K | TS2. 1E202L<br>DL2. E18L<br>MG2. E14L<br>DS2. E202L<br>DA2. S<br>LT. E202L<br>PQ. G (HA)<br>LH. H | FM2. E16L<br>MD2. E18L<br>CF2. E202L<br>ED2. T | CL2. E14L<br>ME2. E202L<br>KJ. E202L<br>KL2. E2G<br>EU2. E21FB | PB2. E202L<br>LB2. E202L<br>FS. E2G | DJ. m<br>Te. y<br>DA. m<br>E02. G (FP) | 82464<br>58<br>HAB#3.0<br>WQ's<br>Group |
| 13:00 | FP. E202L<br>DC. E2G<br>OI. E202L<br>fG. E200<br>DG. A<br>HS. B (PC2, PW)<br>GV. 2E92G | HL. G (PL)<br>DO. E202L<br>DS. G (DF2)<br>D2. B (DQ2)<br>CR. E2G<br>BQ. E2G<br>DO. ME2G<br>LQ. E92G | EX. E3H<br>B2. E2G<br>KQ. E92G<br>AN2. E2G<br>PG. E200<br>QI. E202L<br>KS. E92G | LH. E202L<br>LT. E202L<br>PQ. E92G<br>DA2. E2G<br>MG2. E2G<br>TS2. 1E200 | PC2. G (HS)<br>PW. G (HS)<br>PF. E2G<br>BC2. E14L<br>ED2. E2G<br>MD2. E2G<br>TG2. K<br>AT2. E92G | EU2. E16L<br>MD2. E2G<br>KL2. W<br>DT2. E2G<br>BK. E202L<br>DQ2. G (D2) | FS. E202L<br>BR2. E202L | DF2. B (DS)<br>DA. i<br>PJ2. E202L | 82464<br>47<br>HAB#3.0<br>WQ's<br>Group |
| 13:30 | GV. 2E92G<br>DC. E202L<br>HS. G (HL)<br>Wf. E53L<br>fG. E53L | MY. E202L<br>CR. E53L<br>IE. R<br>LQ. Gs<br>BQ. E2G<br>DO. MK<br>EA. E92G<br>HL. MB (HS)<br>VJ. MT<br>MR. ME2G | KQ. E16L<br>AN2. E18L<br>EX. E92G<br>KS. K<br>PG. E16L<br>QS. E2G<br>IW. T | DA2. G (DO)<br>MG2. T<br>TS2. 1K<br>DS2. E16L<br>AL2. R<br>VP. E92G<br>PQ. E202L | TG2. K<br>AT2. K<br>ED2. E16L<br>FM2. R<br>FN2. E92G<br>VO2. E2G | BK. E92G<br>DQ2. E53L<br>KJ. E92G<br>KL2. E2G<br>DT2. E202L<br>Cf2. K | LB2. E202L<br>BR2. E200<br>FS. H | PJ2. E202L<br>ED2. E14L<br>DA. M<br>PL. M<br>VT. M<br>MS. M | 82464<br>50<br>HAB#3.0<br>WQ's<br>Group |

## RANGING PATTERN

TROOP: Hmu PB's (FDB 6 _RangingPattern) PG: 1\1

MONTH/YEAR: SEPT-2023 OBS: Bun

| DATE | TIME | GRID | θ AC | HAB / VIS | | DISPERSAL | | FOODS | COMMENTS * |
|---|---|---|---|---|---|---|---|---|---|
| | | | | HAB | VIS | Spread | Direct | | |
| 9/14/23 | 6:30 | 0282/215 #0046452 | R | 6.0 | F | S | E | – | ⊖REST |
| | 6:45 | 0282/133 #0046492 | R | 60 | F | S | E | – | " |
| | 7:00 | 0282/136 #0046494 | R | 60 | F | S | E | – | " |
| | 7:15 | 0282/143 #0046507 | R | 60 | F | S | E | – | " |
| | 7:30 | 0282/188 #0046507 | R.T | 60 | F | S | S | – | ⊖ (L) S-SW→S |
| | 7:45 | 0282/359 #0046120 | F.K | 30 | G | S | S | 48F | |
| | 8:00 | 0282/454 #0045686 | F.T.F | 30 | G | S | S | 2G | |
| | 8:15 | 0282/515 #0045506 | F.T.K | 30 | G | S | S | 2G | |
| | 8:30 | 0282/429 #0045261 | F.T.K | 30 | G | S | S | 2G | |
| | 8:45 | 0282/439 #0044473 | F.K | 30 | G | S | S | 2G | |
| | 9:00 | 0282/377 #0044163 | F.T.K | 30 | G | S | S | 2G | |
| | 9:15 | 0282/407 #0043889 | F.T.K | 30 | G | S | S | 2G, 48F | |
| | 9:20 | 0282/411 #0042742 | F.T.F | 30 | G | S | S | 2G | |
| | 9:45 | 0282/490 #0043271 | T.F.K | 30 | G | S | S | 2G | |
| | 10:00 | 0282/396 #0042826 | F.T.K | 30 | G | S | S | 2G | |
| | 10:15 | 0282/339 #0042687 | T.F.K | 30 | F | S | S | 2G | |
| | 10:30 | 0282/275 #0042573 | T.F.K | 30 | F | S | S | 2G | |
| | 10:45 | 0282/096 #0042249 | R.K | 30.5 | F | S | S | – | |
| | 11:00 | 0282/090 #0042113 | R.K | 30.6 | F | S | S | – | |
| | 11:15 | 0282/110 #0042116 | R.K | 30.8 | F | S | S | – | |
| | 11:30 | 0282/181 #0042130 | F.T.F | 30 | F | S | E | 2G | DIR Δ SE |
| | 11:45 | 0282/504 #0042199 | T.F.K | 30 | P | VS | E | 2G | |
| | 12:00 | 0282/620 #0042282 | T.F.K | 30 | P | VS | E | 2G | |
| | 12:15 | 0282/663 #0042242 | F.T.K | 30 | B | VS | E | 2G, 121R | |
| | 12:30 | 0282/697 #0042257 | T.F.K | 30 | B | VS | E | 121R, 2G | |

* In the comments column, note if the troop is at an old bomb site; the cause of a direction change (interference, alarm, etc.) and where applicable, who leads; has a subgroup left the main troop (out of sight); is a consort influencing troop movement, intertroop encounter, etc.

WO's GROUP

R91/2

TROOP: Amu
MONTH/YEAR: AUG / 2022
OBS: BM

# REPRODUCTIVE CYCLE STATES

(B - 15_Reproductivecycle)

10R (above column 12); 10R (above column 26)

| ID | Prev | 1 | 2 | 3 | 4 | 5 | 6 | 7 | 8 | 9 | 10 | 11 | 12 | 13 | 14 | 15 | 16 | 17 | 18 | 19 | 20 | 21 | 22 | 23 | 24 | 25 | 26 | 27 | 28 | 29 | 30 | 31 |
|---|---|---|---|---|---|---|---|---|---|---|---|---|---|---|---|---|---|---|---|---|---|---|---|---|---|---|---|---|---|---|---|---|
| AL | 97 | | 97 | 97 | | | | 95 | 95 | | | | 93 | | | 93 | 93 | 93 | | 92 | | | | | 93 | 92 | 93 | | 93 | 93 | 93 | |
| BQ | 92 | | 92 | 92 | | | | 92 | 92 | | | | 92 | | | 92 | 92 | 92 | | 92 | | | | | 92 | | 92 | | 92 | 92 | 92 | |
| CC | 02 | | 02 | 02 | | | | 88 | 88 | | | | 83 | | | 83 | 83 | 83 | | 83 | | | | | 83 | | 83 | | 83 | 83 | 83 | |
| CR | 03 | | 03 | 03 | | | | 03 | 03 | | | | 03 | | | 03 | 03 | 03 | | 03 | | | | | 03 | | 03 | | 03 | 03 | 03 | |
| DS | 03 | | 03 | 10 | | | | 30 | 30 | | | | 30 | | | 30 | 30 | 30 | | 40+ | | | | | (68) | | 02 | | 02 | 02 | 02 | |
| DO | 86 | | 86 | 86 | | | | 86 | 86 | | | | 86 | | | 86 | 86 | 86 | | 86 | | | | | 87 | | 87 | | 87 | 87 | 87 | |
| DU | 85 | | 85 | 85 | | | | 85 | 85 | | | | 86 | | | 86 | 86 | 86 | | 86 | | | | | 86 | | 86 | | 86 | 86 | 86 | |
| DZ | 02 | | 02 | 02 | | | | 02 | 10 | | | | 20 | | | 20 | 20 | 20 | | 20 | | | | | 20 | | | | 02 | 02 | 02 | |
| EA | 86 | | 86 | 86 | | | | [96] | 96 | | | | 95 | | | 95 | 95 | 95 | | 95 | | | | | 95 | | 95 | | 95 | 95 | 95 | |
| EY | 03 | | △70 | 70 | | | | 70 | 03 | | | | | | | 10 | 10 | 10 | | 20 | | | | | 30 | | 30 | | 30+ | 30+ | 40+ | |
| HA | 02 | | 02 | 02 | | | | 02 | 02 | | | | | | | 02 | 02 | 02 | | 02 | | | | | 10 | | 20 | | 20 | 20 | (68) | |
| HL | 85 | | 85 | 85 | | | | 85 | 83 | | | | | | | 85 | 85 | 85 | | 85 | | | | | 85 | | 85 | | 85 | 85 | 85 | |
| IE | 10 | | 10 | 20 | | | | 30 | 30+ | | | | 40+ | | | (66) | 68 | 02 | | 02 | | | | | 02 | | 02 | | 02 | 02 | 02 | |
| JL | 94 | | 94 | 94 | | | | 94 | 94 | | | | 99 | | | 99 | 99 | 99 | | 99 | | | | | 99 | | 99 | | 99 | 99 | 99 | |
| KT | 40+ | | (64) | 66 | | | | 02 | 02 | | | | 02 | | | 02 | 02 | 02 | | 10 | | | | | 30 | | 30 | | 30+ | 30+ | 30+ | |
| LQ | 93 | | 93 | 93 | | | | 93 | 93 | | | | 92 | | | 92 | 92 | 92 | | 92 | | | | | 92 | | 92 | | 92 | 92 | 92 | |
| MR | 85 | | 85 | 85 | | | | 86 | 86 | | | | 86 | | | 86 | 86 | 86 | | 86 | | | | | 86 | | 86 | | 86 | 86 | 86 | |
| MN | 86 | | 86 | 86 | | | | [96] | 96 | | | | △99 | | | 99 | 99 | 99 | | 99 | | | | | 99 | | 99 | | 99 | 99 | 99 | |
| PY | 30+ | | 30+ | 30+ | | | | 02 | 02 | | | | △70 | | | 70 | 70 | 70 | | 02 | | | | | 02 | | 02 | | 02 | 02 | 02 | |
| TE | 30+ | | (68) | 02 | | | | △70 | 70 | | | | 10 | | | 20 | 20 | 20 | | 20 | | | | | 30 | | 30+ | | (66) | 68 | 02 | |
| TW | 94 | | 94 | 94 | | | | 94 | 94 | | | | | | | 94 | 94 | 94 | | 93 | | | | | 93 | | 93 | | 93 | 93 | 93 | |
| | | | | | | | | | | | | | | | | | | | | | | | | | | | | | | | | |
| | | | | | | | | | | | | | | | | | | | | | | | | | | | | | | | | |
| AL2 | 01 | | 01 | 01 | | | | 01 | 01 | | | | 01 | | | 01 | 01 | 01 | | 01 | | | | | 01 | | 01 | | 01 | 01 | 01 | |
| CF | 84 | | 84 | 84 | | | | 84 | 84 | | | | | | | 84 | 84 | 84 | | 85 | | | | | 85 | | 85 | | 85 | 85 | 85 | |
| DA2 | 01 | | △70 | 70 | | | | 01 | 01 | | | | 10 | | | 10 | 10 | 10 | | 20 | | | | | 20 | | 20 | | 20 | 20 | 20 | |
| DI2 | 84 | | 84 | 84 | | | | 84 | 84 | | | | 85 | | | 85 | 85 | 85 | | 85 | | | | | 85 | | 85 | | 85 | 85 | 85 | |
| DS2 | 30 | | 30 | (62) | | | | 01 | 01 | | | | | | | 01 | △70 | 70 | | 20 | | | | | 30 | | 30 | | 30 | 30 | 30 | |
| LH | 20 | | 20 | 30 | | | | 01 | 01 | | | | 10 | | | △70 | 70 | 72 | | 72 | | | | | 20 | | 20 | | 20 | 20 | 20 | |
| LT | 70 | | 10 | 20 | | | | 30 | 30 | | | | 30 | | | 30+ | 30+ | 30 | | (66) | | | | | 01 | | 01 | | 01 | 01 | 01 | |
| LY | 86 | | 86 | 86 | | | | 86 | 86 | | | | 87 | | | 87 | 87 | 87 | | 87 | | | | | 87 | | 87 | | 87 | 87 | 87 | |
| MG2 | 30 | | 30 | 30 | | | | 30 | 30 | | | | (68) | | | 01 | 01 | 01 | | 01 | | | | | [72] | | 72 | | 72 | 20 | 20 | |
| PQ | 70 | | 70 | 70 | | | | 72 | 72 | | | | 30 | | | 30 | 30 | 30 | | 30 | | | | | (66) | | 66 | | 01 | 01 | 01 | |

COMMENTS: (Births, Promotions, Unusual events of cycle - swelling/bleeding during pregnancy, Disease)

CC Promoted From c-fc-Pres A♀ In this month of Aug 2023

EA was Promoted From Pres- Lact A♀ on 8/5/23

VI Promoted from Pres. Lact A♀ on 8/18/23

MN Promoted From Pres- Lact A♀ on 8/6/23

○ The First Day of Deflation.
□ The Day of Birth
△ The First Day of Menstruation

A♀'s (left margin beside first group); SA♀'s (left margin beside second group)

Original Design by SCS

Modified: 9/09

255

PREPARED BY SCS

PREDATION SUMMARY DECEMBER 1978

GILGIL BABOON PROJECT

PG

| DATE | TIME START | TIME STOP | CAPTURE SEEN | TYPE OF CAPTURE | PREY SPECIES/AGE | LOC. | COVER | CONSUMPTION CARCASS | CONSUMPTION SCRAPS | CONSUMPTION ATTEND | COMMENTS | OBS |
|---|---|---|---|---|---|---|---|---|---|---|---|---|
| 12/7 | 10:20 | ? | yes SH | accidental discovery & lunge | HARE / baby | P29 | bush | RD ? | ~~[illegible]~~ ? | BO AG SH | SH gets hare - it screams - voc brings BO RD AG ~ SH RD ~ SH for 2 min before she drops it [illegible] RD | SCS |
| 12/8 | ? 1:00 | ? | no didn't follow | — | TOMMY / neonatal | P24 | open | AG ? | | CH GA RD EC subgroup | on edge of during long distance interloop & AG crosses in flood & CH + closely EC animals interested & follow also | |
| 12/12 | 10:16 - 10:30 | | no | — | CHICKEN / adult | P23 | house | GA DU | AG TH CY HT | | middle of - Africans ~ PHG - lots of agg around chicken | |
| 12/23 | 11:50 | - 1:05 + | yes RD | accidental discovery & lunge | TOMMY / neonatal | P24 | open | RD MQ TB... | BS GA MQ CH CL TB RD DU HI TE | middle of but all attendants get meat | see notes - incredible polyadic agg - also RD ~~fearful~~ but ~~tenaciously~~ holds onto c | |

Appendix 3

# Troop Movement Factors

The following needs to be recorded on the data sheet:

1. Primary Foods Targeted (see food list)
2. Secondary Food Targeted
3. Interference
   a. By another troop
   b. By local people
   c. By livestock alone
   d. By livestock with herder
   e. By alarm: real or false
   f. By consort group movements
4. Negotiation
   a. Indicator animal leaving
   b. Subgrouping due to ecological reasons
   c. Subgrouping due to social reasons aside from consorts
   d. Getting lost
   e. Male-male agonism/aggression
   f. Positive interactions with another troop like following
5. Resource Movements
   a. Exploration of new areas (going on "safari")
   b. Searching for water
   c. Discovering a preferred food while foraging

6. Type of Change in Troop Movement Direction
   a. Reaction to any of the above
   b. Adjustment to any of the above
   c. Reorientation to any of the above
   d. Reinforcement to any of the above
   e. Don't know

Appendix 4

# Fusion Criteria

1. *Proximity*: the way the troop is mixed should tell us about whether they are fused. When they are fused completely, we might expect the nearest neighbors to be members of either troop (after taking kin proximity into account). This is in contrast to the clustering of one troop within or on the edge/boundary of the combined troop. The degree of mixing may be different at different times of day, but a "fused" troop will show mixed animals throughout the day.
2. *Grooming*: grooming between members of the two troops (crossover grooming).
3. *Other social (not play or greeting)*: greeting and play is common in intertroop interactions and in early stages of fusion. Considering the types of social interactions that go on within a normal troop, do these also occur? Look for friendships between females of the two troops or between males/infants or males/females that cut across the troops as an indicator of fusion.
4. *Foraging*: the distribution during foraging is a sign of fusion. The two troops being consistently mixed during foraging rather than foraging separately or creating troop subgroups during foraging indicates fusion.

5. *Dynamics during the day (movement and feeding)*: we have to think about the dynamics of the two troops during the day. This means
   a. at the sleeping site
   b. leaving the sleeping site
   c. orientation during troop movement
   d. monitoring each other's position
   e. coming together and moving apart
   f. returning to the rocks
   g. sleeping at the rocks

In order to decide about "fusion," we need to think about how this is different or similar from one troop dividing into subgroups for resources (ecological subgrouping). When there is subgrouping, it can't be based on previous troop membership, and we expect the subgroups to be mixed.

6. *Social dynamics*: there are various social interactions that take place within a troop which are different between troops. These can be useful criteria to decide about fusion
   a. competition over resources (foods, estrous females, new infants)
   b. possessiveness over social partners (male/female, females/infants, male/infant)
   c. support of friends and allies (crossover alliances)
   d. lack of vigilance of one troop by the other
7. *Dynamics of interactions with other troops*: this may not be a good criterion for deciding on fusion because when the two troops were separate, they often allied with each other to displace or harass a third troop. Do they respond as one troop rather than in a three-way intertroop interaction?
8. *Integrated female dominance hierarchy*
9. *Integrated male dominance hierarchy*

10. *Possessiveness of females in the presence of local troops or new immigrant males*: possessiveness extends to all females, not just home troop females (crossover herding). This is like the stages of adolescent male transfers who first actively herd all females as if demonstrating their allegiance and membership and self-identification.
11. *Previously vigilant of the other troop vs. now monitoring of the mixed troop to maintain social cohesion*
    a. reaction to separation (monitoring and wahoos)
    b. reaction to reunion (reunion embraces and grunt rounds)
12. *Crossover troop movement negotiations* (rather than one troop responding to moves by the other troop) ID animal can be from either troop and can get response from members of the opposite troop.
13. *Crossover mobbing*: as when a male frightens any female or infant and animals from both troops respond.
14. *Crossover participation in predation events*
15. *Fusion of home range* (like MLK going into STT extended home range and beyond their own home range boundaries)

*Note*: Initially we also listed crossover consorts, and this took a long time to develop for the MLK male WG (Wiggle). However, males from outside troops do join a troop for the sole purpose of consorts (see HK in early days of MLK; see examples later for STT/MLK and for PHG/BRD and PHG/MMR). Thus, although this should be monitored, it is not a very valuable criterion on its own.

We classified MLK/STT as fused based on these criteria in July 2001, and starting August 2001, NABO (NBO)—meaning "coming together" in Maa—was created.

**Stages of Fusion**

This refers to the criteria listed above.

Stage 1: female affiliation, no tension, female interaction, integrated female hierarchy (basically "female-based" bonding)
Stage 2: male-female interaction and affiliation (male-female bonding)
Stage 3: integrated male dominance hierarchy (male-male interactions as within one troop: crossover consorts with follower from both troops)

Appendix 5

# Zilla/Heather Genealogy

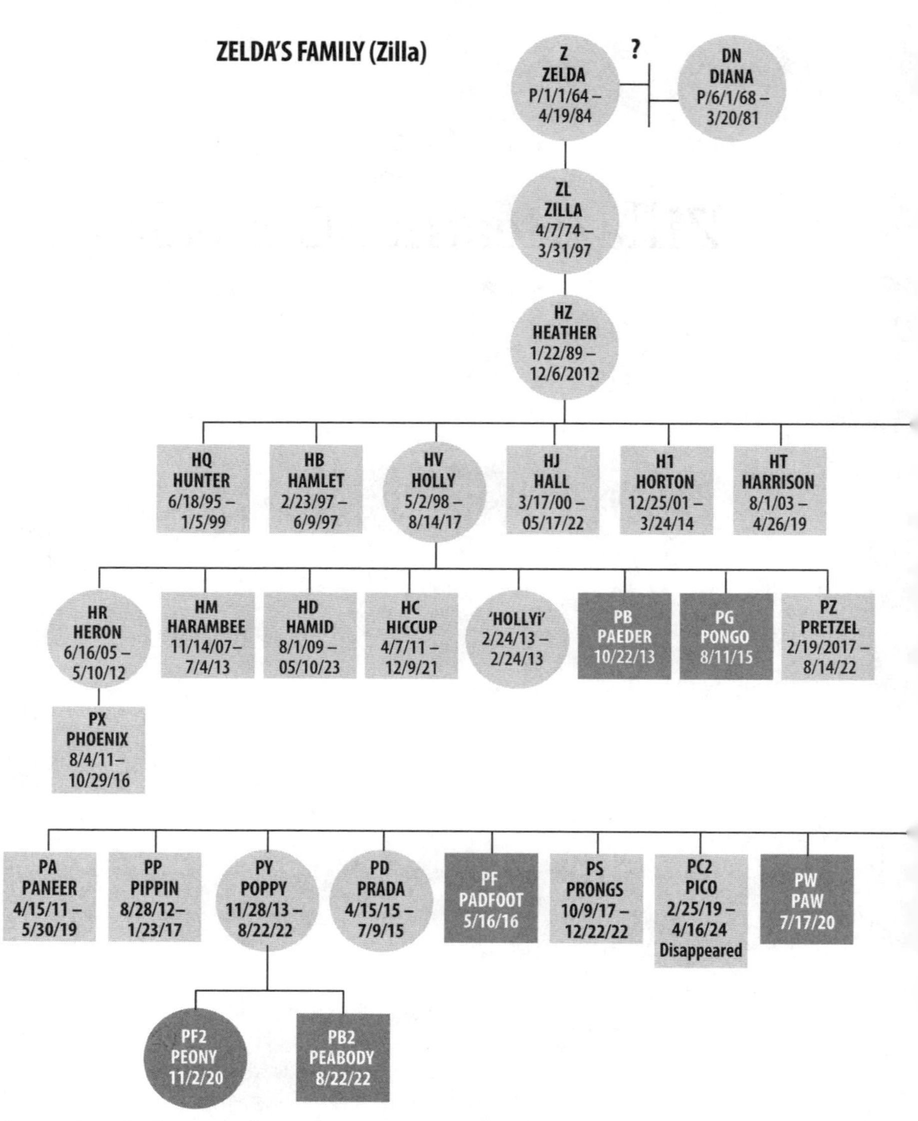


The dark gray color is for those living at the time of submission.

→ **Diana's Family**

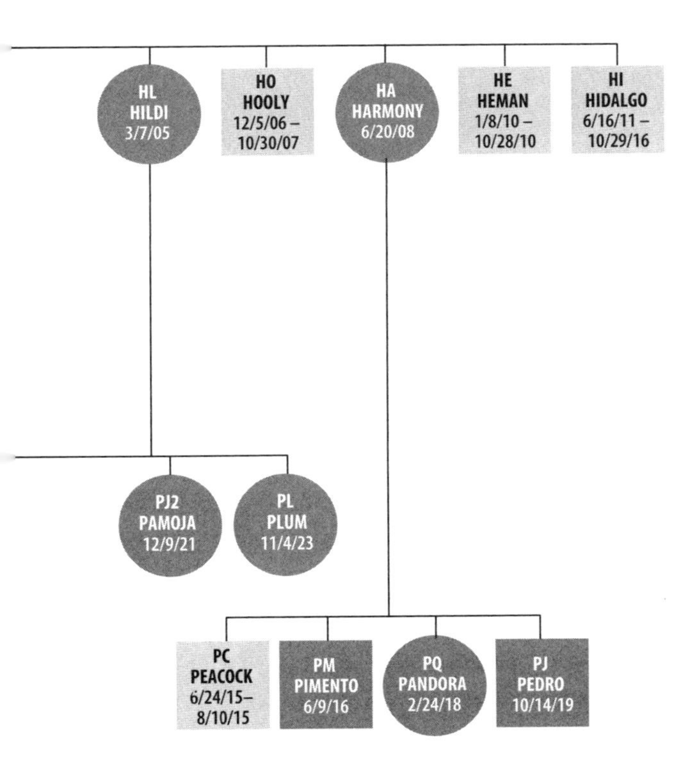

Appendix 6

# Troop History

# TROOP HISTORY

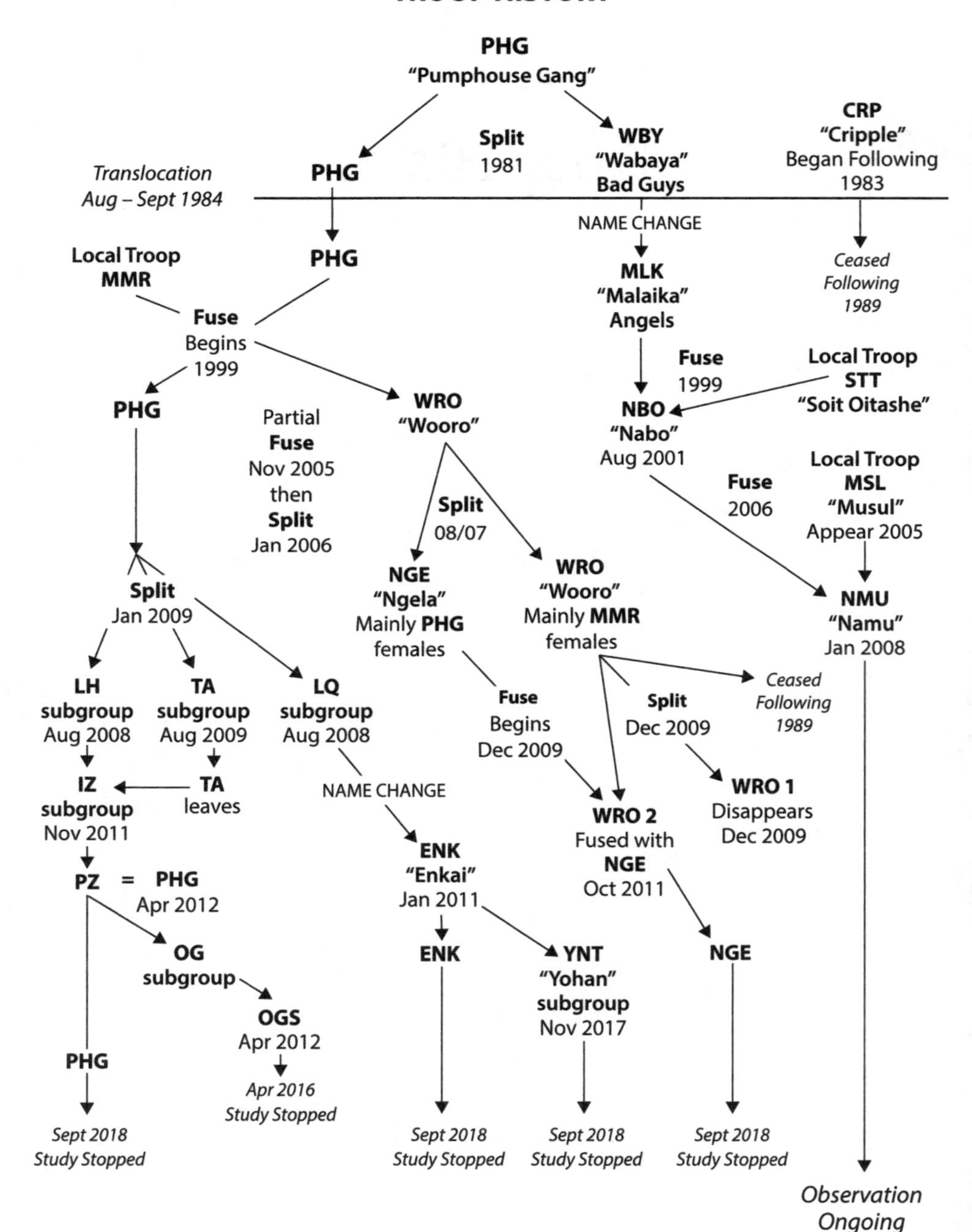

Appendix 7

# The Problem with the Cape Baboons

## Sent to the *Cape Times*, Cape Town, South Africa for Circulation

*Dr. Shirley C. Strum, July 21, 2012*

I have studied baboons in Kenya for 40 years, including working on baboon-human conflict. In addition to my behavioral research, I also specialize in other conservation issues. Part of my expertise focuses on the development of the ethics behind conservation and animal welfare, and I have taught a very successful course called Conservation and the Human Predicament at the University of California since the 1980s. Because of these skills, I have been an advisor on many conservation projects. In the last 12 months, I visited Cape Town twice as a consultant on baboon-human conflict and to compare notes with researchers and managers in an attempt to understand and help resolve the problematic relationship between humans and baboons there.

All the reading that I had done in advance of my visits to the Cape Baboons did not prepare me for what I witnessed firsthand in the

Cape. In February of this year, I visited the majority of the troops around Cape Point and urban areas. The troops that I saw had taken baboon ingenuity and adaptability to their logical extreme. Most have reached the point where they are uncontrollable, either because they have become dominant to the people living in the area or to the monitors trying to control them, and/or because the topography of some of the locations makes it extremely difficult to control them.

I place the blame on the extreme position of baboon activists who have thwarted the only methods that might have prevented this situation. Baboons, in their relations with each other and, by extension, in their relations with humans, need to know the boundaries and limits of acceptable behavior. Appropriate behavior is created not just by benefits but by threats about misbehavior; these expectations need to be constantly reinforced. My extensive research on food-enhanced baboons (supported by studies from elsewhere) demonstrates the great "benefit" that human food has for baboons and how it changes their behavior, diet, and activity budget, making it even easier for them to wait patiently for the opportunity to raid human foods. There are good evolutionary reasons that motivate baboons once they get a taste of the higher-quality and larger packages of human foods. I have tested and witnessed many control techniques in the last 40 years. They all point to the need to make the costs of raiding greater than the benefits baboons get from this behavior.

The baboons should have been aversively deterred from approaching and feeding on human food from the start and had this deterrence maintained consistently. How could anyone let a troop sleep on the roof of an apartment building? It is a joke to have monitors walking behind the baboons and simply clapping hands. At this point, I'm not even certain that major deterrent efforts will be effective for many of the troops, but it is the only option now, short of eliminating most or all of the baboons. The epitaph of these baboons will read: "Met an untimely end because activists could not face reality."

I am scandalized by the publicity campaign mounted by the activists. I care about baboons as much or more than they do. But because I care, I would sacrifice some to save the whole if that is what it takes. By contrast, activists seem to only care about how they (humans) feel about baboons. They seem unable to take the baboons' point of view or get out of their particular (and in the context of the rest of Africa, peculiar) ethnocentric opinions about animal welfare. The current baboon approach advocated by "activists" is unsustainable and in my mind, unconscionable. This is made worse by the clear hypocrisy of the outcry of some of the same people who practiced deterrence in the past. It would have worked if it was done well, consistently, and on a continuous basis. Sadly, it wasn't.

If there is to be a trial for the "murder" of specific baboon males, I would testify as an expert witness that the very same people bringing the charges are the ones who should be on trial because it was their objections that prevented methods that could have saved these baboons. The future of the Cape Baboons is being endangered by the very people shouting the loudest against the only appropriate methods we have now. If deterrence had been used successfully earlier, there would be no need to kill any baboons today. The only good outcome of these untoward attacks is that they stimulated the Baboon Research Unit (BRU) to study what options remain. The BRU has used the best science in the service of conservation. They have thought outside of the box. They should be commended, not condemned. I cannot speak about the other agencies involved, but this "anti-science" stance of the activists demonstrates the ignorance that opinion matters as much as evidence.

Sadly, as with many conservation issues, it boils down to human foibles: matters of control or money or visibility/credit which play out regardless of whether conservation actually happens.

I strongly urge the activists to stop this senseless campaign. Instead, they should use that energy to help support the reasonable efforts that are being proposed. If they don't, they will have more baboon deaths on their conscience, and I won't forget that.

Dr. Shirley C. Strum
Director, Uaso Ngiro Baboon Project
and
Professor of Anthropology
University of California, San Diego
La Jolla, CA 92093-0532
U.S.A.

# Notes

## *Preface*

1. Kenneth M. Weiss and Anne V. Buchanan, *The Mermaid's Tale: Four Billion Years of Cooperation in the Making of Living Things* (Cambridge, MA: Harvard University Press, 2009); Massimo Pigliucci and Gerd B. Muller, eds., *Evolution: The Extended Synthesis* (Cambridge, MA: MIT Press, 2010); Itai Yanai and Martin Lercher, *The Society of Genes* (Cambridge, MA: Harvard University Press, 2016); and Alfonso Martinez Arias, *The Master Builder: How the New Science of the Cell Is Rewriting the Story of Life* (New York: Basic Books, 2023).

## *Chapter 1. Why Baboons? (1972)*

1. K. Ron L. Hall, "Behaviour and Ecology of the Wild Patas Monkey, *Erythrocebus patas*, in Uganda," *Journal of Zoology*, no. 148 (1965): 15–87; J. Steve Gartlan, "Adaptive Aspects of Social Structure in *Erythrocebus patas*," in *Proceedings of the Fifth Congress of the International Primatological Society*, ed. S. Kondo, M. Kawai, A. Ehara, and S. Kawamura (Tokyo: Japanese Science Press, 1975), 161–71.

2. Sherwood L. Washburn and Irven DeVore, "Social Behavior of Baboons and Early Man," in *Social Life of Early Man*, ed. Sherwood L. Washburn (Chicago: Aldine, 1961), 91–105.

3. Sherwood L. Washburn, "The Analysis of Primate Evolution with Particular Reference to the Origin of Man," *Cold Spring Harbor Symposia on Quantitative Biology* 15 (1951): 67–78.

4. Jane Goodall, *In the Shadow of Man* (Boston: Houghton Mifflin, 1971).

5. Sherwood L. Washburn and Irven DeVore, "Baboon Ecology and Human Evolution," in *African Ecology and Human Evolution*, ed. F. Clark Howell and Francois Bourliere (New York: Wenner-Gren Foundation for Anthropological Research, 1963), 335–67; Irven DeVore and K. R. L. Hall, "Baboon Ecology," in *Primate Behavior: Field Studies of Monkeys and Apes*, ed. Irven DeVore (New York: Holt, Rinehart and Winston, 1965), 20–52.

6. K. R. L. Hall and Irven DeVore, "Baboon Social Behavior," in DeVore, *Primate Behavior*, 53–110.

7. Irven DeVore and K. R. L. Hall, "Baboon Ecology and Human Evolution," in Howell and Bourliere, *African Ecology*, 335–67.

8. Thelma Rowell, "Forest Living Baboons in Uganda," *Journal of Zoology* 149 (1966): 344–64.

9. Timothy W. Ransom, *Beach Troop of the Gombe* (Lewisburg, PA: Bucknell University Press, 1981).

10. Raymond J. Rhine, "The Order of Movement of Yellow Baboons (*Papio cynocephalus*)," *Folia Primatologica* 23 (1975): 72–104; Robert S. O. Harding, "Patterns of Movement in Open Country Baboons," *American Journal of Physical Anthropology* 47 (1977): 349–53, https://doi.org/10.1002/ajpa.1330470215.

11. B. Latour and S. C. Strum, "Human Social Origins: Oh Please, Tell Us Another Story," *Journal of Social and Biological Systems* 9, no. 2 (1986): 168–87, https://doi.org/10.1016/0140-1750(86)90027-8.

### *Chapter 2. Learning Baboon (1972–1973)*

1. K. Imanishi, "Social Organization of Subhuman Primates in Their Natural Habitat," *Current Anthropology* 1 (1960): 399–407; J. Itani, "Twenty Years with Mount Takasaki Monkeys," in *Primate Utilization and Conservation*, ed. G. Bermant and D. G. Lindburg (New York: John Wiley and Sons, 1975), 101–25; Naoki Koyama, "On Dominance Rank and Kinship of a Wild Japanese Monkey Troop in Arashiyama," *Primates* 8 (1967): 189–216.

2. Jane Goodall, *In the Shadow of Man* (Boston: Houghton Mifflin, 1971).

### *Chapter 3. The Model Breaks (1972–1981)*

1. Susan C. Alberts, Heather E. Watts, and Jeanne Altmann, "Queing and Queue-Jumping: Long-Term Patterns of Reproductive Skew in Male Savannah Baboons, *Papio cynocephalus*," *Animal Behavior* 65 (2003): 821–40, https://doi.org/10.1006/anbe.2003.2106.

2. Timothy W. Ransom, *Beach Troop of the Gombe* (Lewisburg, PA: Bucknell University Press, 1981).

3. A. G. Hendrickx and D. C. Kraemer, "Observations of the Menstrual Cycle, Optimal Mating Time and Pre-Implantation Embryos of the Baboon (*Papio anubis* and *Papio cynocephalus*)," *Journal of Reproduction and Fertility* 6 (1969): 119–28.

4. Mark Kirkpatrick, "Sexual Selection by Female Choice in Polygynous Animals," *Annual Review of Ecology and Systematics* 18 (1987): 43–70.

5. David Sloan Wilson, "The Group Selection Controversy, History and Current Status," *Annual Review of Ecology and Systematics* 14 (1983): 159–87.

6. Shirley C. Strum, "Why Males Use Infants," in *Primate Paternalism*, ed. David Taub (New York: Van Nostrand Reinhold, 1983), 146–85.

7. Ransom, *Beach Troop of the Gombe*.

8. Robert C. Solomon and Fernando Flores, *Building Trust in Business, Politics, Relationships, and Life* (New York: Oxford University Press, 2003).

9. Michele J. Gelfand et al., "Differences between Tight and Loose Cultures: A 33-Nation Study," *Science* 332 (2011): 1100–1104.

10. Shirley C. Strum, "Agonistic Dominance in Male Baboons: An Alternative View," *International Journal of Primatology* 3, no. 2 (1982), https://doi.org/10.1007/BF02693494.

11. Shirley C. Strum, *Almost Human: A Journey into the World of Baboons* (New York: Random House, 1987).

12. Glenn Hausfater, *Dominance and Reproduction in Baboons (Papio cynocephalus)*, ed. H. Kuhn, W. P. Luckett, and C. R. Noback, *Contributions to Primatology*, vol. 7 (Basel: S. Karger, 1975).

13. Leigh L. Thompson, Jiunwen Wang, and Brian C. Gunia, "Negotiation," *Annual Review of Psychology* 61 (2010): 491–515.

14. Oliver E. Williamson, "Transaction Cost Economics: The Natural Progression," *American Economic Review* 100 (2010): 673–90, https://doi.org/10.1257/aer.100.3.673.

15. Shirley C. Strum, "Use of Females by Male Olive Baboons," *American Journal of Primatology* 5 (1983): 93–109.

16. Joan B. Silk et al., "Strong and Consistent Social Bonds Enhance the Longevity of Female Baboons," *Current Biology* 20 (2010): 1–3, https://doi.org/10.1016/j.cub.2010.05.067; Elizabeth C. Lange et al., "Early Life Adversity and Adult Social Relationships Have Independent Effects on Survival in a Wild Primate," *Science Advances* 9, no. 20 (2023), https://doi.org/10.1126/sciadv.ade7172.

17. Robert D. Putnam, *Bowling Alone: The Collapse and Revival of American Community* (New York: Simon and Schuster, 2000).

18. Lawrence Nolan, ed., *The Cambridge Descartes Lexicon* (Cambridge: Cambridge University Press, 2015).

19. Strum, *Almost Human*.

20. Charles Darwin, *The Descent of Man, and Selection in Relation to Sex* (1871; reprint, Princeton, NJ: Princeton University Press, 1981).

21. Shirley C. Strum, "Primate Predation: Interim Report on the Development of a Tradition in a Troop of Olive Baboons," *Science* 187, no. 4178 (1975), https://doi.org/10.1126/science.187.4178.755.

22. Shirley C. Strum, "Life with the Pumphouse Gang: New Insights into Baboon Behavior," *National Geographic Magazine* 147 (1975): 762–91.

### *Chapter 4. Bitter Harvest (1981–1984)*

1. Shirley C. Strum, *Almost Human: A Journey into the World of Baboons* (New York: Random House, 1987).

2. David Western, "The Biodiversity Crisis: A Challenge for Biology," *Oikos* 63 (1992): 29–38, https://doi.org/10.2307/3545513.

3. Shirley C. Strum, "Primate Predation: Interim Report on the Development of a Tradition in a Troop of Olive Baboons," *Science* 187, no. 4178 (1975): 755–57, https://doi.org/10.1126/science.187.4178.755.

4. John Mitani and Peter S. Rodman, "Territoriality: The Relation of Ranging Patterns and Home Range Size to Defendibility with an Analysis of Territoriality among Primate Species," *Behavioral Ecology and Sociobiology* 5 (1979): 241–51; Claudia Kasper and Bernhard Voelkl, "A Social Network Analysis of Primate Groups," *Primates* 50 (2009): 343–56.

5. Shirley C. Strum, "Agonistic Dominance in Male Baboons: An Alternative View," *International Journal of Primatology* 3, no. 2 (1982): 175–202, https://doi.org/10.1007/BF02693494.

6. Robert M. Eley, Shirley C. Strum, Gerald Muchemi, and Graham D. F. Reid, "Nutrition, Body Condition, Activity Patterns, and Parasitism of Free-Ranging Troops of Olive Baboons (*Papio anubis*) in Kenya," *American Journal of Primatology* 18, no. 3 (1989): 209–19.

7. Karl von Frisch, "Decoding the Language of the Bee," *Science* 185, no. 4152 (1974): 663–68; Konrad Lorenz, "The Comparative Method in Studying Innate Behavior Patterns," *Symposia of the Society for Experimental Biology* 4 (1950): 221–68; Niko Tinbergen, *The Study of Instinct* (New York: Oxford University Press, 1974).

8. Charles Darwin, *On the Origin of Species by Means of Natural Selection* (London: J. Murray, 1859).

9. Irven DeVore, *Primate Behavior: Field Studies of Monkeys and Apes* (New York: Holt, Rinehart and Winston, 1965); Phyllis Jay, *Primates: Studies in Adaptation and Variability* (New York: Holt, Rinehart, and Winston, 1968).

10. John H. Crook and Steve Gartlan, "Evolution of Primate Societies," *Nature* 210 (1966): 1200–1203.

11. T. T. Struhsaker, "Correlates of Ecology and Social Organization among African Cercopithecines," *Folia Primatologica* 11, no. 1 (1969): 80–118.

12. E. O. Wilson, *Sociobiology: The New Synthesis* (Cambridge, MA: Harvard University Press, 1975).

### *Chapter 5. The People Problem (1981–1984)*

1. Debra L. Forthman, Shirley C. Strum, and Gerald M. Muchemi, "Applied Conditioned Taste Aversion and the Management and Conservation of Crop-Raiding Primates," in *Commensalism and Conflict: The Human-Primate Interface*, ed. J. D. Paterson and Janette Wallis (Norman, OK: American Society of Primatologists, 2005), 420–43.

2. John Mitani and Peter S. Rodman, "Territoriality: The Relation of Ranging Patterns and Home Range Size to Defendibility with an Analysis of Territoriality among Primate Species," *Behavioral Ecology and Sociobiology* 5 (1979): 241–51; Claudia Kasper and Bernhard Voelkl, "A Social Network Analysis of Primate Groups," *Primates* 50 (2009): 343–56.

3. David Western, Michael Wright, and Shirley C. Strum, *Natural Connection: Perspectives in Community-Based Conservation* (Washington, DC: Island Press, 1994).

4. "Human Misery as Drought, Hunger Ravage Counties," *Sunday Nation* (Kenya), 1997.

5. "Starving Boy Dead after Dog Meat Meal," *Sunday Nation* (Kenya), 1997.

6. David M. Amodio, "The Neuroscience of Prejudice and Stereotyping," *Nature Reviews Neuroscience* 15 (2014): 670–82, https://doi.org/10.1038/nrn3800.

7. Daniel M. Wegner and Kurt Gray, *The Mind Club: Who Thinks, What Feels, and Why It Matters* (New York: Viking, 2016); Daniel M. Wegner, "How to Think, Say, or Do Precisely the Worst Thing for Any Occasion," *Science* 325, no. 5936 (2009): 48–50, https://doi.org/10.1126/science.1167346; Joseph Henrich and Michael Muthukrishna, "The Origins and Psychology of Human Cooperation," *Annual Review of Psychology* 72 (2021): 207–40, https://doi.org/10.1146/annurev-psych-081920-042106.

## *Chapter 6. Rehomed (1983–1984)*

1. Shirley C. Strum, "The Development of Primate Raiding: Implications for Management and Conservation," *International Journal of Primatology* 31, no. 1 (2010): 133–56, https://doi.org/10.1007/s10764-009-9387-5.

2. Shirley C. Strum, *Almost Human: A Journey into the World of Baboons* (New York: Random House, 1987).

3. Shirley C. Strum and Charles H. Southwick, "Translocation of Primates," in *Primates: The Road to Self-Sustaining Populations*, ed. Kurt Benirschke (New York: Springer-Verlag, 1987), 949–58.

4. S. C. Strum, "Measuring Success in Primate Translocation: A Baboon Case Study," *American Journal of Primatology* 65, no. 2 (2005): 117–40, https://doi.org/10.1002/ajp.20103.

## *Chapter 7. Strangers in a Strange Land (1984–1988)*

1. Shirley C. Strum, "The Development of Primate Raiding: Implications for Management and Conservation," *International Journal of Primatology* 31, no. 1 (2010): 133–56, https://doi.org/10.1007/s10764-009-9387-5.

2. Diane B. Paul, "The Selection of the 'Survival of the Fittest,'" *Journal of the History of Biology* 21 (1988): 411–24; Herbert Spenser, *Principles of Biology*, vol. 1 (London: Williams and Norgate, 1864).

3. Charles Darwin, *On the Origin of Species by Means of Natural Selection* (London: J. Murray, 1859); Alfred Russel Wallace, "On the Physical Geography of the Malay Archipelago," *Journal of the Royal Geographical Society of London* 33 (1863): 217–34.

4. Richard Dawkins, *The Selfish Gene* (Oxford: Oxford University Press, 1976).

5. Thomas Malthus, *An Essay on the Principle of Population or a View of Its Past and Present Effect on Human Happiness*, vol. 1 (London: J. Johnson, 1807).

6. Sherwood L. Washburn, "The New Physical Anthropology," *Transactions of the New York Acadmic of Sciences, Series II* 13, no. 7 (1951): 298–304; Shirley C. Strum, Donald G. Lindburg, and David Hamburg, *The New Physical Anthropology:*

*Science, Humanism, and Critical Reflection* (Upper Saddle River, NJ: Prentice Hall, 1999).

7. Charles Darwin, *The Descent of Man and Selection in Relation to Sex* (1871; reprint, Princeton, NJ: Princeton University Press, 1981).

8. Shirley C. Strum, "Measuring Success in Primate Translocation: A Baboon Case Study," *American Journal of Primatology* 65, no. 2 (2005): 117–40, https://doi.org/10.1002/ajp.20103.

9. Antoine de Saint-Exupery, *The Little Prince* (San Diego: Harcourt Brace Jovanovich, 1941).

### *Chapter 8. Troop Movements (1986–2000)*

1. Andrew Berdahl, Colin J. Torney, Christos C. Ioannou, Jolyon J. Faria, and Iain D. Couzin, "Emergent Sensing of Complex Environments by Mobile Animal Groups," *Science* 339, no. 6119 (2013): 574–76, https://doi.org/10.1126/science.1225883.

2. M. Mitchell Waldrop, *Complexity: The Emerging Science at the Edge of Order and Chaos* (New York: Touchstone/Simon and Schuster, 1992); Neil F. Johnson, *Simply Complexity: A Clear Guide to Complexity Theory* (Oxford: One World, 2009).

3. Ruth Parkin-Gounelas, *The Psychology and Politics of the Collective: Groups, Crowds, and Mass Identification* (London: Routledge, 2012).

4. Ariana Strandburg-Peshkin, Damien R. Farine, Iain D. Couzin, and Margaret C. Crofoot, "Shared Decision-Making Drives Collective Movement in Wild Baboons," *Science* 348, no. 6241 (2015): 1358–61, https://doi.org/10.1126/science.aaa5099.

5. Hans Kummer, *In Quest of the Sacred Baboon: A Scientist's Journey* (Princeton, NJ: Princeton Unversity Press, 1995).

6. Liana Y. Zanette and Michael Clinchy, "Ecology of Fear," *Current Biology* 29, no. 9 (2019): PR309–R313, https://doi.org/10.1016/j.cub.2019.02.042.

7. Shirley C. Strum, Deborah Forster, and Edwin Hutchins, "Why Machiavellian Intelligence May Not Be Machiavellian," in *Machiavellian Intelligence, II: Extensions and Evaluations*, ed. Andrew Whiten and Richard Byrne (New York: Cambridge University Press, 1997), 50–85.

8. Jacques Monod, *Chance and Necessity* (New York: Knopf, 1971).

### *Chapter 9. Mergers and Acquisitions (1999–2001)*

1. Bruno Latour, *Science in Action* (Cambridge, MA: Harvard University Press, 1987); Bruno Latour, *Reassembling the Social: An Introduction to Actor-Network Theory* (Oxford: Oxford University Press, 2005).

2. Stuart A. Altmann and Jeanne Altmann, *Baboon Ecology: African Field Research* (Chicago: University of Chicago Press, 1970); S. K. Wasser, G. W. Norton, R. J. Rhine, N. Klein, and S. Kleindorfer, "Aging and Social Rank Effects on the Reproductive System of Free-Ranging Yellow Baboons (*Papio cynocephalus*) at Mikumi National Park, Tanzania," *Human Reproduction Update* 4, no. 4 (1998): 430–38.

3. Marc D. Hauser, Dorothy L. Cheney, and Robert M. Seyfarth, “Group Extinction and Fusion in Free-Ranging Vervet Monkeys,” *American Journal of Primatology* 11 (1986): 63–77; L. A. Isbell, Dorothy L. Cheney, and Robert M. Seyfarth, “Group Fusions and Minimum Group Sizes in Vervet Monkeys (*Cercopithecus aethiops*),” *American Journal of Primatology* 25 (1991): 57–65; Karin Enstam Jaffe and Lynne A. Isbell, “Changes in Ranging and Agonistic Behavior of Vervet Monkeys (*Cercopithecus aethiops*) after Predator-Induced Group Fusion,” *Amercian Journal of Primatology* 72 (2010): 634–44; and Wolfgang P. J. Dittus, “Group Fusion among Wild Toque Macaques: An Extreme Case of Intergroup Resource Competition,” *Behaviour* 100 (1987): 247–91.

4. Julia Lehmann, Amanda H. Korstjens, and R. I. M. Dunbar, “Fission-Fusion Social Systems as a Strategy for Coping with Ecological Constraints: A Primate Case,” *Evolutionary Ecology* 21 (2007): 613–34, https://doi.org/10.1007/s10682-006-9141-9.

5. Assaf Anyamba, Compton J. Tucker, and Robert Mahoney, “From El Niño to La Niña: Vegetation Response Patterns over East and Southern Africa during the 1997–2000 Period,” *Journal of Climate* 15 (2002): 3096–3103, https://doi.org/10.1175/1520-0442(2002)015<3096:FENOTL>2.0.CO;2.

6. Peter M. Kappeler and Joan B. Silk, eds., *Mind the Gap: Tracing the Origins of Human Universals* (New York: Springer, 2010).

### *Chapter 10. The Power of Predictability (1972–2008)*

1. E. H. M. Sterk, David P. Watts, and Carel P. Van Schaik, “The Evolution of Female Social Relationships in Nonhuman Primates,” *Behavioral Ecology and Sociobiology* 41 (1997): 291–309; Shuichi Matsumura, “Evolution of ‘Egalitarian’ and ‘Despotic’ Social Systems among Macaques,” *Primates* 40, no. 1 (1999): 23–31.

2. Konrad Lorenz, *On Aggression* (London: Meuthen, 1966).

3. W. D. Hamilton, “The Moulding of Senescence by Natural Selection,” *Journal of Theoretical Biology* 12 (1966): 12–45.

4. E. O. Wilson, *Sociobiology: The New Synthesis* (Cambridge, MA: Harvard University Press, 1975).

5. Jeanne Altmann, *Baboon Mothers and Infants* (Cambridge, MA: Harvard University Press, 1980).

6. Wilson, *Sociobiology*.

7. Sarah Hrdy, *The Woman Who Never Evolved* (Cambridge MA: Harvard University Press, 1999).

8. Glenn Hausfater, *Dominance and Reproduction in Baboons (Papio cynocephalus)*, ed. H. Kuhn, W. P. Luckett, and C. R. Noback (Basel: S. Karger, 1975); Jeanne Altmann, Glenn Hausfater, and Stuart A. Altmann, “Determinants of Reproductive Success in Savannah Baboons,” in *Reproductive Success*, ed. Timothy Clutton-Brock (Chicago: University of Chicago Press, 1988), 403–18; Dorothy L. Cheney et al., “Reproduction, Mortality and Female Reproductive Success in Chacma Baboons of the Okavango Delta, Botswana,” in *Reproduction and Fitness*

*in Baboons: Behavior, Ecology and Life History Perspectives*, ed. Larissa Swedell and Steven R. Leigh (New York: Springer, 2006), 147–76.

9. Shirley C. Strum and Jonah David Western, "Variations in Fecundity with Age and Environment in a Baboon Population," *American Journal of Primatology* 3 (1982): 67–76.

10. Joan B. Silk et al., "Strong and Consistent Social Bonds Enhance the Longevity of Female Baboons," *Current Biology* 20, no. 15 (2010): 1–3, https://doi.org/10.1016/j.cub.2010.05.067; Joan B. Silk, Susan C. Alberts, and Jeanne Altmann, "Social Bonds of Female Baboons Enhance Infant Survival," *Science* 302, no. 5648 (2003): 1231–34, https://doi.org/10.1126/science.1088580; Jonah D. Western and Shirley C. Strum, "Sex, Kinship, and the Evolution of Social Manipulation," *Ethology and Sociobiology* 4, no. 1 (1983): 19–28, https://doi.org/10.1016/0162-3095(83)90004-3; S. K. Wasser, G. W. Norton, S. Kleindorfer, and R. J. Rhine, "Population Trend Alters the Effect of Maternal Dominance Rank on Lifetime Reproductive Success in Yellow Baboons (*Papio cynocephalus*)," *Behavioral Ecology and Sociobiology* 56 (2004): 338–45; Amanda J. Lea et al., "Dominance Rank-Associated Gene Expression Is Widespread, Sex-Specific, and a Precursor to High Social Status in Wild Male Baboons," *PNAS* 115, no. 52 (2018): E12163–71, https://doi.org/10.1073/pnas.1811967115.

11. Gregory S. Berns, Jonathan Chappelow, Milos Cekic, Caroline F. Zink, Guiseppe Pagnoni, and Megan E. Martin-Skurski, "Neurobiological Substrates of Dread," *Science* 312, no. 5774 (2006): 754–58, https://doi.org/10.1126/science.1123721; Justin S. Feinstein Ralph Adolphs, Antonio Damasio, and Daniel Tranel, "The Human Amygdala and the Induction and Experience of Fear," *Current Biology* 21, no. 1 (2011): 34–38, https://doi.org/10.1016/j.cub.2010.11.042.

12. Caroline F. Zink, Yunxia Tong, Qiang Chen, Danielle S. Bassett, Jason L. Stein, and Andreas Meyer-Lindenberg, "Know Your Place: Neural Processing of Social Hierarchy in Humans," *Neuron* 58, no. 2 (2008): 273–83, https://doi.org/10.1016/j.neuron.2008.01.025.

13. Jae-Young Son, Apoorva Bhanderi, and Oriel Feldman-Hall, "Abstract Cognitive Maps of Social Network Structure Aid Adaptive Inference," *PNAS* 120, no. 47 (2023): e2310801120, https://doi.org/10.1073/pnas.2310801120.

14. Roman M. Wittig, Catherine Crockford, Julia Lehmann, Patricia L. Whitten, Robert M. Seyfarth, and Dorothy L. Cheney, "Focused Grooming Networks and Stress Alleviation in Wild Female Baboons," *Hormones and Behavior* 54, no. 1 (2008): 170–77, https://doi.org/10.1016/j.yhbeh.2008.02.009; Catherine Crockford, Roman M. Wittig, Patricia L. Whitten, Robert M. Seyfarth, and Dorothy L. Cheney, "Social Stressors and Coping Mechanisms in Wild Female Baboons (*Papio hamadryas ursinus*)," *Hormones and Behavior* 53, no. 1 (2008): 254–65, https://doi.org/10.1016/j.yhbeh.2007.10.007; Roman Wittig, Catherine Crockford, Anja Weltring, Kevin E. Langergraber, Tobias Deschner, and Klaus Zuberbühler, "Social Support Reduces Stress Hormone Levels in Wild Chimpanzees across Stressful Events and Everyday Affiliations," *Nature Communications* 7, no. 13361 (2016), https://doi.org/10.1038/ncomms13361.

15. Susan C. Alberts, Robert M. Sapolsky, and Jeanne Altmann, "Behavioral, Endocrine, and Immunological Correlates of Immigration by an Aggressive Male into a Natural Primate Group," *Hormones and Behavior* 26, no. 2 (1992): 167–78, https://doi.org/10.1016/0018-506X(92)90040-3; Lea et al., "Dominance Rank-Associated Gene Expression."

16. Laurence R. Gesquire, Niki H. Learn, Carolina M. Simao, Patrick O. Onyango, Susan C. Alberts, and Jeanne Altmann, "Life at the Top: Rank and Stress in Wild Male Baboons," *Science* 333, no. 6040 (2011): 357–60, https://doi.org/10.1126/science.1207120.

17. Sam K. Patterson, Katie Hinde, Angela B. Bond, Benjamin C. Trumble, Shirley C. Strum, and Joan B. Silk, "Effects of Early Life Adversity on Maternal Effort and Glucocorticoids in Wild Olive Baboons," *Behavioral Ecology and Sociobiology* 75, no. 114 (2021): 114, https://doi.org/10.1007/s00265-021-03056-7.

18. Francis Fukuyama, *The Origins of Political Order: From Prehuman Times to the French Revolution* (New York: Farrar, Straus, and Giroux, 2011).

### *Chapter 11. Mob Story (2007–2012)*

1. CABI International, "*Opuntia stricta* (erect prickly pear)," November 27, 2007, https://doi.org/10.1079/cabicompendium.37728.

2. Shirley C. Strum, Graham Stirling, and Steve Kalusi Mutunga, "The Perfect Storm: Land Use Change Promotes *Opuntia stricta*'s Invasion of Pastoral Rangelands in Kenya," *Journal of Arid Environments* 118 (2015): 37–47, https://doi.org/10.1016/j.jaridenv.2015.02.015.

3. Llewellyn C. Foxcroft, Mathieu Rouget, David M. Richardson, and Sandra Mac Fadyen, "Reconstructing 50 Years of *Opuntia stricta* Invasion in the Kruger National Park, South Africa: Environmental Determinants and Propogule Pressure," *Diversity and Distributions* 10, no. 5–6 (2004): 427–37, https://doi.org/10.1111/j.1366-9516.2004.00117.x.

4. CABI International, "*Opuntia stricta.*"

5. Catherine Kunyanga and J. K. Imungi, "Nutrient Contents of the Opuntia Cactus Fruit, Syrup and Leaves/Pads" (University of Nairobi Report, 2009).

6. Strum, Stirling, and Mutunga, "The Perfect Storm."

### *Chapter 12. Finding Meaning in a Mistake (2012–2015)*

1. Scott L. Brincat, Markus Siegel, Constantin von Nicolai, and Earl K. Miller, "Gradual Progression from Sensory to Task-Related Processing in Cerebral Cortex," *PNAS* 115, no. 30 (2018): e-7202–11, https://doi.org/10.1073/pnas.1717075115; Robert M. Sapolsky, *Behave: The Biology of Humans at Our Best and Worst* (New York: Penguin, 2017).

2. Anthony D. Barnosky et al., "Has the Earth's Sixth Mass Extinction Already Arrived?," *Nature* 471 (2011): 51–57, https://doi.org/10.1038/nature09678.

3. Stephen L. Brusatte et al., "The Extinction of the Dinosaurs," *Biological Reviews* 90, no. 2 (2014): 628–42, https://doi.org/10.1111/brv.12128; Dan Ferber, "Orangutans Face Extinction in the Wild," *Science* 288, no. 5469 (2000): 1147–49,

https://doi.org/10.1126/science.288.5469.1147b; Chih-Ming Hung et al., "Drastic Population Fluctuations Explain the Rapid Extinction of the Passenger Pigeon," *PNAS* 111, no. 29 (2014): 10636–41, https://doi.org/10.1073/pnas.1401526111.

4. Shirley C. Strum, "Science Encounters," in *Primate Encounters: Models of Science, Gender and Society*, ed. Shirley C. Strum and Linda M. Fedigan (Chicago: University of Chicago Press, 2000).

5. S. J. Gould and R. C. Lewontin, "The Spandrels of San Marco and the Panglossian Paradigm: A Critique of the Adaptationist Programme," *Proceedings of the Royal Society B* 205 (1979): 581–98, https://doi.org/10.1098/rspb.1979.0086; K. N. Laland and G. R. Brown, *Sense and Nonsense: Evolutionary Perspective on Human Behavior* (New York: Oxford University Press, 2011).

6. Dipesh Chakrabarty, "Anthropocene Time," *History and Theory* 57, no. 1 (2018): 5–32, https://doi.org/10.1111/hith.12044.

7. Peter M.Vitousek, Harold A. Mooney, Jane Lubchenco, and Jerry M. Melillo, "Human Domination of Earth's Ecosystems," *Science* 277, no. 5325 (1997): 494–99, https://doi.org/10.1126/science.277.5325.494.

8. Shirley C. Strum, "Why Natural History Is Important to (Primate) Science: A Baboon Case Study," *International Journal of Primatology* 40 (2019): 596–612, https://doi.org/10.1007/s10764-019-00117-7.

9. Shirley C. Strum, Deborah Forster, and Edwin Hutchins, "Why Machiavellian Intelligence May Not Be Machiavellian," in *Machiavellian Intelligence, II: Extensions and Evaluations*, ed. Andrew Whiten and Richard Byrne (New York: Cambridge University Press, 1997), 50–85.

10. Kenneth M. Weiss and Anne V. Buchanan, *The Mermaid's Tale: Four Billion Years of Cooperation in the Making of Living Things* (Cambridge, MA: Harvard University Press, 2009).

11. Weiss and Buchanan, *The Mermaid's Tale*, 230.

## *Chapter 13. Group Think (2008–2018)*

1. C. P. van Schaik and J. A. R. A. M. van Hoof, "Why Are Diurnal Primates Living in Groups?," *Behaviour* 87 (1983): 120–44; E. H. M. Sterk, David P. Watts, and Carel P. Van Schaik, "The Evolution of Female Social Relationships in Nonhuman Primates," *Behavioral Ecology and Sociobiology* 41 (1997): 291–309.

2. R. W. Wrangham, "An Ecological Model of Female-Bonded Primate Groups," *Behaviour* 75 (1980): 335–69.

3. Shirley C. Strum, Graham Stirling, and Steve Kalusi Mutunga, "The Perfect Storm: Land Use Change Promotes *Opuntia stricta*'s Invasion of Pastoral Rangelands in Kenya," *Journal of Arid Environments* 118 (2015): 37–47, https://doi.org/10.1016/j.jaridenv.2015.02.015.

4. S. S. Strum and Bruno Latour, "Redefining the Social Link: From Baboons to Humans," *Social Science Information* 26, no. 4 (1987): 783–802, https://doi.org/10.1177/053901887026004004.

5. R. I. M. Dunbar, "The Social Brain Hypothesis," *Evolutionary Anthropology* 6 (1998): 178–90; R. A. Barton and R. I. M. Dunbar, "Evolution of the Social Brain,"

in *Machiavellian Intelligence, II: Extensions and Evaluations*, ed. Andrew Whiten and Richard Byrne (New York: Cambridge University Press, 1997), 240–88.

6. Katharine Milton, "Distribution Patterns of Tropical Plant Foods as a Stimulus to Primate Mental Development," *American Anthropologist* 83 (1981): 534–48; Alex R. DeCasien, Scott A. Williams, and James P. Higham, "Primate Brain Size Is Predicted by Diet but Not Sociality," *Nature Ecology and Evolution* 1, no. 0112 (2017), https://doi.org/10.1038/s41559-017-0112.

7. Kate Teffer and Katerina Semendeferi, "Human Prefrontal Cortex: Evolution, Development, and Pathology," in *Progress in Brain Research*, ed. Michael A. Hofman and Dean Falk (New York: Elsevier, 2012), 191–218, https://doi.org/10.1016/B978-0-444-53860-4.00009-x; Robert M. Sapolsky, *Behave: The Biology of Humans at Our Best and Worst* (New York: Penguin, 2017), 45–50.

8. Robin I. M. Dunbar, "The Social Brain Meets Neuroimaging," *Trends in Cognitive Science* 16, no. 2 (2012): 101–2, https://doi.org/10.1016/j.tics.2011.11.013.

9. Kei Watanabe and Shintaro Funahasi, "Neural Mechanisms of Dual-Task Interference and Cognitive Capacity Limitation on the Prefrontal Cortex," *Nature Neuroscience* 17 (2014): 601–11, https://doi.org/10.1038/nn.3667.

10. C. Nathan Dewall, Roy F. Baumeister, Matthew T. Gailliot, and Jon K. Maner, "Depletion Makes the Heart Grow Less Helpful: Helping as a Function of Self-Regulatory Energy and Genetic Relatedness," *Personality and Social Psychology Bulletin* 34, no. 12 (2008): 1653–62, https://doi.org/10.1177/0146720832398l.

11. Federica Amici, Filippo Aureli, and Josep Call, "Fission-Fusion Dynamics, Behavioral Flexibility, and Inhibitory Control in Primates," *Current Biology* 18, no. 18 (2008): 1415–19, https://doi.org/10.1016/j.cub.2008.08.020.

12. J. Sallet et al., "Social Network Size Affects Neural Circuits in Macaques," *Science* 334, no. 6056 (2011): 697–700, https://doi.org/10.1126/science.1210027.

13. Susan C. Alberts, Robert M. Sapolsky, and Jeanne Altmann, "Behavioral, Endocrine, and Immunological Correlates of Immigration by an Aggressive Male into a Natural Primate Group," *Hormones and Behavior* 26, no. 2 (1992): 167–78, https://doi.org/10.1146/annurev.anthro.33.070203.144000; T. J. Bergman, J. C. Beehner, D. L. Cheney, R. M. Seyfarth, and P. L. Whitten, "Correlates of Stress in Free-Ranging Male Chacma Baboons, *Papio hamadryas ursinus*," *Animal Behavior* 70, no. 3 (2005): 703–13, https://doi.org/10.1016/j.anbehav.2004.12.017.

14. Roman M. Wittig, Catherine Crockford, Anja Weltring, Kevin E. Langergraber, Tobias Deschner, and Klaus Zuberbühler, "Social Support Reduces Stress Hormone Levels in Wild Chimpanzees across Stressful Events and Everyday Affiliations," *Nature Communications* 7, no. 13361 (2016), https://doi.org/10.1038/ncomms13361; Bergman et al., "Correlates of Stress"; Sarah D. Carnegie, Linda M. Fedigan, and Toni E. Ziegler, "Social and Environmental Factors Affecting Fecal Glucocoricoids in Wild, Female White-Faced Capuchins (*Cebus capucinus*)," *American Journal of Primatology* 73, no. 9 (2011): 861–69, https://doi.org/10.1002/ajp.20954; Evelyn L. Pain, Andreas Koenig, Anthony Di Fiore, and Amy Lu, "Behavioral and Physiological Responses to Instability in Group

Membership in Wild Male Woolly Monkeys (*Lagothrix lagotricha loeppigii*)," *American Journal of Primatology* 83, no. 3 (2021): e23240, https://doi.org/10.1002/ajp.23240.

15. Robert D. Putnam, *Bowling Alone: The Collapse and Revival of American Community* (New York: Simon and Schuster, 2000).

### *Chapter 14. Why Baboons Are Not Human: The Matter of Mind*

1. Edwin Hutchins, *Cognition in the Wild* (Cambridge, MA: MIT Press, 1995); Gavriel Salomon, *Distributed Cognitions: Psychological and Educational Considerations* (Cambridge: Cambridge University Press, 1997).

2. Robert A. Wilson, "Collective Memory, Group Minds, and the Extended Mind Thesis," *Cognitive Processes* 6 (2005): 227–36, https://doi.org/10.1007/s10339-005-0012-z.

3. Salomon, *Distributed Cognitions*, xiii.

4. Shirley C. Strum, Deborah Forster, and Edwin Hutchins, "Why Machiavellian Intelligence May Not Be Machiavellian," in *Machiavellian Intelligence, II: Extensions and Evaluations*, ed. Andrew Whiten and Richard Byrne (New York: Cambridge University Press, 1997), 50–85.

5. Hutchins, *Cognition in the Wild*, 225.

6. Strum, Forster, and Hutchins, "Why Machiavellian Intelligence May Not Be Machiavellian."

7. L. S. Vygotsky, *Mind in Society: The Development of Higher Psychological Processes* (Cambridge, MA: Harvard University Press, 1978); James V. Wertsch, *Vygotsky and the Social Formation of Mind* (Cambridge, MA: Harvard University Press, 1985).

8. Lucy Suchman, *Plans and Situated Actions: The Problem of Human-Machine Communication* (New York: Cambridge University Press, 1987); Barbara Rogoff, *Apprenticeship in Thinking: Cognitive Development in Social Context* (New York: Oxford University Press, 1990); Barbara Rogoff, *The Cultural Nature of Human Development* (New York: Oxford University Press, 2003).

9. Anthony Formaux, Dan Sperber, Joël Fagot, and Nicolas Claidière, "Guinea Baboons Are Strategic Cooperators," *Science Advances* 9, no. 43 (2023), https://doi.org/10.1126/sciadv.adi5282.

10. Federica Amici, Filippo Aureli, and Josep Call, "Monkeys and Apes: Are Their Cognitive Skills Really so Different?," *American Journal of Physical Anthropology* 143, no. 2 (2010): 188–97, https://doi.org/10.1002/ajpa.21305.

11. Daniel Dennett, "Intentional Systems," *Journal of Philosophy* 68, no. 4 (1971): 87–106.

12. David Premack and Guy Woodruff, "Does the Chimpanzee Have a Theory of Mind?," *Behavioral and Brain Sciences* 1, no. 4 (1978): 515–26.

13. On fish, see Masanori Kohda et al., "If a Fish Can Pass the Mark Test, What Are the Implications for Consciousness and Self-Awareness Testing in Animals?," *PLoS Biology* 17, no. 2 (2019): e3000021. On mice, see Jun Yokose,

William D. Marks, and Takashi Kitamura, "Visuotactile Integration Facilitates Mirror-Induced Self-Directed Behavior through Activation of Hippocampal Neuronal Ensembles in Mice," *Neuron* 112, no. 2 (2024): 306–18, https://doi.org/10.1016/j.neuron.2023.10.022.

14. Atsushi Iriki and Osamu Sakura, "The Neuroscience of Primate Intellectual Evolution: Natural Selection and Passive and Intentional Niche Construction," *Philosophical Transactions of the Royal Society B: Biological Sciences* 363, no. 1500 (2008): 2229–41.

15. Claudio Tennie, Josep Call, and Michael Tomasello, "Ratcheting up the Rachet: On the Evolution of Cumulative Culture," *Philosophical Transactions of the Royal Society B: Biological Sciences* 364 (2009): 2405–15, https://doi.org/10.1098/rstb.2009.0052.

16. Shirley C. Strum and Bruno Latour, "Redefining the Social Link: From Baboons to Humans," *Social Science Information* 26, no. 4 (1987): 783–802, https://doi.org/10.1177/053901887026004004.

17. David Premack and Ann James Premack, *The Mind of an Ape* (New York: W. W. Norton, 1983); R. Allen Gardener and Beatrice T. Gardener, "Teaching Sign Language to a Chimpanzee: A Standardized System of Gestures Provides a Means of Two-Way Communication with a Chimpanzee," *Science* 165 (1969): 664–72; Herbert Terrace, *Nim: A Chimpanzee Who Learned Sign Language* (New York: Penguin Random House, 1980); Francine Patterson and Eugene Linden, *The Education of Koko* (New York: Holt, Rinehart, and Winston, 1981); Sue Savage-Rumbaugh and Roger Lewin, *Kanzi: The Ape at the Brink of the Human Mind* (New York: Wiley, 1994).

18. Savage-Rumbaugh and Lewin, *Kanzi*.

19. Erica A. Cartmill, "Overcoming Bias in the Comparison of Human Language and Animal Communication," *PNAS* 120, no. 47 (2023): e2218799120, https://doi.org/10.1073/pnas.2218799120.

20. Cartmill, "Overcoming Bias."

21. Richard Dawkins, "An Open Letter to Prince Charles," *Edge*, May 20, 2000, https://www.edge.org/conversation/richard_dawkins-an-open-letter-to-prince-charles.

22. Martin E. P. Seligman, Peter Railton, Roy F. Baumeister, and Chandra Sripada, *Homo Prospectus* (Oxford: Oxford University Press, 2016).

23. Chet C. Sherwood and Aida Gómez-Robles, "Brain Plasticity and Human Evolution," *Annual Review of Anthropology* 46 (2017): 399–419, https://doi.org/10.1146/annurev-anthro-102215-100009.

### *Chapter 15. Why Baboons Are Not Human: Culture and Evolution*

1. Bruno Latour and Steve Woolgar, *Laboratory Life: The Social Construction of Scientific Factors* (Princeton, NJ: Princeton University Press, 1979).

2. Erving Goffman, *The Performance of Self in Everyday Life* (Garden City, NY: Doubleday, 1959).

3. Shirley C. Strum and Bruno Latour, "Redefining the Social Link: From Baboons to Humans," *Social Science Information* 26, no. 4 (1987): 783–802, https://doi.org/10.1177/053901887026004004.

4. Bruno Latour, *Reassembling the Social: An Introduction to Actor-Network Theory* (Oxford: Oxford University Press, 2005).

5. Strum and Latour, "Redefining the Social Link."

6. J. Itani and A. Nishimura, "The Study of Infrahuman Culture in Japan," in *Precultural Primate Behavior*, ed. Emil W. Menzel, vol. 1 (Basel: Karger, 1973), 26–55

7. A. Whiten et al., "Culture in Chimpanzees," *Nature* 399 (1999): 682–85.

8. Carel P. Van Schaik et al., "Orangutan Cultures and the Evolution of Material Culture," *Science* 299, no. 5603 (2003): 102–5.

9. Tiago Falótico, Tomos Proffitt, Eduaro B. Ottoni, Richard A. Staff, and Michael Haslam, "Three Thousand Years of Wild Capuchin Tool Use," *Nature Ecology and Evolution* 3 (2019): 1034–38, https://doi.org/10.1038/s41559-019-0904-4.

10. Michael Chimento, Gustavo Alarcon-Nieto, and Lucy Aplin, "Immigrant Birds Learn from Socially Observed Differences in Payoffs When Their Environment Changes," *PLoS Biology* 22, no. 11 (2024): e3002699, https://doi.org/10.1371/journal.pbio.3002699.

11. Elena Kerjean, Erica van de Waal, and Charlotte Canteloup, "Social Dynamics of Vervet Monkey Groups Are Dependent upon Group Identity," *IScience* 27, no. 1 (2023): 108591, https://doi.org/10.1016/j.isci.2023.108591.

12. Claudio Tennie, Josep Call, and Michael Tomasello, "Ratcheting up the Rachet: On the Evolution of Cumulative Culture," *Philosophical Transactions of the Royal Society B: Biological Sciences* 364 (2009): 2405–15, https://doi.org/10.1098/rstb.2009.0052; Andrew Whiten, Dora Biro, Nicolas Bredeche, Ellen C. Garland, and Simon Kirby, "The Emergence of Collective Knowledge and Cumulative Culture in Animals, Humans and Machines," *Philosophical Transactions of the Royal Society B: Biological Sciences* 377 (2021): e20200306, https://doi.org/10.1098/rstb.2020.0306.

13. Robert M. Sapolsky and Lisa J. Share, "A Pacific Culture among Wild Baboons: Its Emergence and Transmission," *PLoS Biology* 2, no. 4 (2004): 534–41, https://doi.org/10.1371/journal.pbio.0020106.

14. Chimento, Alarcon-Nieto, and Aplin, "Immigrant Birds Learn from Socially Observed Differences."

15. Philip Ball, *How Life Works: A User's Guide to the New Biology* (Chicago: University of Chicago Press, 2023).

16. Charles Darwin, *On the Origin of Species by Means of Natural Selection* (London: J. Murray, 1859; see particularly the section "Mutual Affinities of Organic Beings"); Rui Diogo, "Etho-Eco-Morphological Mismatches, an Overlooked Phenomenon in Ecology, Evolution, and Evo-Devo That Supports ONCE (Organic Nonoptimal Constrained Evolution) and the Key Evolutionary Role of

Organismal Behavior," *Frontiers in Ecology and Evolution* 5 (2017), https://doi.org/10.3389/fevo.2017.00003.

17. Peter M. Kappeler and Joan B. Silk, eds., *Mind the Gap: Tracing the Origins of Human Universals* (New York: Springer, 2010), https://doi.org/10.1007/978-3-642-02725-3.

18. Jacques Monod, *Chance and Necessity* (New York: Knopf, 1971).

19. Joseph Henrich and Michael Muthukrishna, "The Origins and Psychology of Human Cooperation," *Annual Review of Psychology* 72 (2021): 207–40, https://doi.org/10.1146/annurev-psych-081920-042106.

20. David Western, *We Alone: How Humans Have Conquered the Planet and Can Also Save It* (New Haven, CT: Yale University Press, 2020).

21. Frans de Waal, *Are We Smart Enough to Know How Smart Animals Are?* (New York: W. W. Norton, 2016).

## *Chapter 16. Vindication*

1. Irwin S. Bernstein, "Metaphor, Cognitive Belief and Science; Commentary on Tactical Deception in Primates by A. Whiten and R. W. Byrne," *Behavioral and Brain Sciences* 11, no. 2 (1988): 247–48.

2. Joan Didion, "Planting a Tree Is Not a Way of Life," commencement speech, University of California Riverside, 1975, quoted in J. D. Warren, "Joan Didion's 'Lost' Commencement Address, Revealed," *UC Riverside News*, January 10, 2022, https://news.ucr.edu/articles/2022/01/10/joan-didions-lost-commencement-address-reveals.

3. D. L. Forthman-Quick, "Controlling Primate Pests: The Feasibility of Conditioned Taste Aversion," in *Current Perspectives in Primate Social Dynamics*, ed. D. M. Taub and F. A. King (New York: Van Nostrand Reinhold, 1986), 252–73; Debra L. Forthman, Shirley C. Strum, and Gerald M. Muchemi, "Applied Conditioned Taste Aversion and the Management and Conservation of Crop-Raiding Primates," in *Commensalism and Conflict: The Human-Primate Interface*, ed. J. D. Paterson and Janette Wallis (Norman, OK: American Society of Primatologists, 2005), 420–43.

4. John Herman Randall, "The Development of Scientific Method in the School of Padua," *Journal of the History of Ideas* 1, no. 2 (1940): 177–206, www.jstor.org/stable/2707332.

5. Charles Lyell and Gérard Paul Deshayes, *Principles of Geology: Being an Attempt to Explain the Former Changes of the Earth's Surface, by Reference to Causes Now in Operation*, vol. 1 (London: J. Murray, 1830); Alexander von Humboldt and Aime Bonpland, *Personal Narrative of Travels to the Equinoctial Regions of the New Continent, During the Years 1799–1804*, vol. 5 (London: Longman, Hurst, Rees, Orme, and Brown, 1827); Alfred Russel Wallace, "On the Physical Geography of the Malay Archipelago," *Journal of the Royal Geographical Society of London* 33 (1863): 217–34; Charles Darwin, *On the Origin of Species by Means of Natural Selection* (London: J. Murray, 1859); Iain McCalman, *Darwin's Armada: Four*

*Voyagers to the Southern Oceans and Their Battle for the Theory of Evolution* (London: Pocket Books, 2009).

6. Randall, "The Development of Scientific Method."

7. Karl Popper, *The Logic of Scientific Discovery* (Vienna, 1935; repr., London: Routledge, 1992).

8. John R. Platt, "Strong Inference," *Science* 146, no. (3642) (1964): 347–53.

9. See Irwin S. Bernstein, "Metaphor, Cognitive Belief and Science; Commentary on Tactical Deception in Primates by A. Whiten and R. W. Byrne," *Behavioral and Brain Sciences* 11, no. 2 (1988): 247–48.

10. Charles S. Elton, *The Ecology of Invasions by Plants and Animals* (London: Methuen, 1958).

11. Paul K. Dayton, "The Importance of Natural Sciences to Conservation," *The American Naturalist* 162, no. 1 (2003): 1–13; Joshua J. Tewksbury et al., "Natural History's Place in Science and Society," *BioScience* 64, no. 4 (2014): 300–310, https://doi.org/10.1093/biosci/biu032; Harry W. Greene and Jonathan B. Losos, "Systematics, Natural History, and Conservation: Field Biologists Must Fight a Public-Image Problem," *BioScience* 38, no. 7 (1988): 458–62, https://www.jstor.org/stable/1310949; G. E. Belovsky et al., "Ten Suggestions to Strengthen the Science of Ecology," *BioScience* 54, no. 4 (2004): 345–51; Rafe Sangarin and Anibal Pauchard, *Observation and Ecology: Broadening the Scope of Science to Understand a Complex World* (Washington, DC: Island Press, 2012).

12. Kerwin Lee Klein, "In Search of Narrative Mastery: Postmodernism and the People without History," *History and Theory* 34, no. 4 (1995): 275–98, www.jstor.org/stable/2505403.

13. Jonathan Gottschall, *The Storytelling Animal: How Stories Make Us Human* (Boston: Mariner Books, 2013); Lucas B. Bietti, Ottlie Tilson, and Adrian Bangerter, "Storytelling as Adaptive Collective Sensemaking," *Topics in Cognitive Science* 11 (2019): 710–32, https://doi.org/10.1111/tops.12358.

14. Kenneth M. Weiss and Anne V. Buchanan, *The Mermaid's Tale: Four Billion Years of Cooperation in the Making of Living Things* (Cambridge, MA: Harvard University Press, 2009), 61.

15. Weiss and Buchanan, *The Mermaid's Tale*, 64.

16. Dipesh Chakrabarty, "Anthropocene Time," *History and Theory* 57 (2018): 5–32, https://doi.org/10.1111/hith.12044.

17. Charles Darwin, *The Descent of Man, and Selection in Relation to Sex* (1871; reprint, Princeton, NJ: Princeton University Press, 1981).

### *Chapter 17. Science in the Wild*

1. Iain McCalman, *Darwin's Armada: Four Voyages to the Southern Ocean and Their Battle for the Theory of Evolution* (London: Pocket Books, 2009).

2. Charles Darwin, *The Voyage of the Beagle* (1839; reprint, New York: Penguin Books, 1989).

## Chapter 18. Interpretations

1. Shirley C. Strum, "Agonistic Dominance in Male Baboons: An Alternative View," *International Journal of Primatology* 3, no. 2 (1982), https://doi.org/10.1007/BF02693494.

2. Susan C. Alberts, Heather E. Watts, and Jeanne Altmann, "Queing and Queue-Jumping: Long-Term Patterns of Reproductive Skew in Male Savannah Baboons, *Papio cynocephalus*," *Animal Behavior* 65, no. 4 (2003): 821–40, https://doi.org/10.1006/anbe.2003.2106.

3. Frans de Waal, *Are We Smart Enough to Know How Smart Animals Are?* (New York: W. W. Norton, 2016); Frans de Waal, *Mama's Last Hug: Animal Emotions and What They Tell Us about Ourselves* (New York: W. W. Norton, 2019).

4. Evelyn Fox Keller, *Reflections on Gender and Science* (New Haven, CT: Yale University Press, 1996).

5. Hope Jahren, *Lab Girl: A Memoir* (New Haven, CT: Knopf Doubleday, 2017), 75, 76.

6. Linda M. Fedigan, *Primate Paradigms: Sex Roles and Social Bonds* (Montreal: Eden Press, 1982); Meredith Small, *Female Choices: Sexual Behavior of Female Primates* (Ithaca, NY: Cornell University Press, 1993); Sarah Hrdy, *The Woman Who Never Evolved* (Cambridge, MA: Harvard University Press, 1999); Alison Jolly, *Lucy's Legacy: Sex and Intelligence in Human Evolution* (Cambridge, MA: Harvard University Press, 1999).

7. Sandra Harding, *The Science Question in Feminism* (Ithaca, NY: Cornell Unversity Press, 1986); Keller, *Reflections on Gender and Science*; Donna Haraway, "Monkeys, Aliens, and Women: Love, Science, and Politics at the Intersection of Feminist Theory and Colonial Discourse," *Women's Studies International Forum* 12, no. 3 (1989): 295–312.

8. Shirley C. Strum and Linda Marie Fedigan, *Primate Encounters: Models of Science, Gender, and Society* (Chicago: University of Chicago Press, 2000).

9. Bruno Latour, *Science in Action* (Cambridge, MA: Harvard University Press, 1987).

10. Strum and Fedigan, *Primate Encounters*.

11. Bruno Latour and Shirley S. Strum, "Human Social Origins: Oh Please, Tell Us Another Story," *Journal of Social and Biological Systems* 9, no. 2 (1986): 169–87, https://doi.org/10.1016/0140-1750(86)90027-8.

12. Harry F. Harlow and Margaret Harlow, "Learning to Love," *American Scientist* 54, no. 3 (1966): 244–72, www.jstor.org/stable/27836477.

## Chapter 19. Forces of Nature

1. Roderick Frazier Nash, *Wilderness and the American Mind* (New Haven, CT: Yale University Press, 2014).

2. Keith Thomas, *Man and the Natural World: Changing Attitudes in England 1500–1800* (Oxford: Oxford University Press, 1996).

3. Roderick Frazier Nash, *The Rights of Nature* (New Haven, CT: Yale University Press, 1989).

4. Nash, *Wilderness and the American Mind.*

5. David Western, *In the Dust of Kilimanjaro* (Washington DC: Shearwater Island Press, 1997).

6. Paul J. Crutzen, "The Anthropocene," in *Earth System Science in the Anthropocene*, ed. Eckart Ehlers and Thomas Krafft (Berlin: Springer, 2006), 13–18.

7. Paola Cavalieri and Peter Singer, *The Great Ape Project: Equality beyond Humanity* (New York: St. Martin's Griffin, 1994).

8. Paola Cavalieri, "The Meaning of the Great Ape Project," *Politics and Animals* 1 (2015): 16–34.

9. Nash, *The Rights of Nature.*

10. Nash, *Wilderness and the American Mind.*

11. Marian Stamp Dawkins, "Animal Welfare and Effective Farming: Is Conflict Inevitable?," *Animal Production Science* 57, no. 2 (2016): 201–8, https://doi.org/10.1071/AN15383; Temple Grandin, *Temple Grandin's Guide to Working with Farm Animals: Safe, Humane Livestock Handling Practices for the Small Farm* (New York: Storey Publishing/Hachette Book Group, 2017).

12. C. Josh Donlan et al., "Pleistocene Rewilding: An Optomistic Agenda for Twenty-First Century Conservation," *American Naturalist* 168, no. 5 (2006): 660–81, https://doi.org/10.1086/508027.

13. David Western, *We Alone: How Humans Have Conquered the Planet and Can Also Save It* (New Haven, CT: Yale University Press, 2020).

14. Lee Alan Dugatkin, *The Altruism Equation: Seven Scientists Search for the Origins of Goodness* (Princeton, NJ: Princeton University Press, 2006).

### *Chapter 20. Coexistence in a World in Flux*

1. Valentina Fiasco and Kati Massarella, "Human-Wildlife Coexistence: Business as Usual Conservation or an Opportunity for Transformative Change?," *Conservation and Society* 20, no. 2 (2022): 167–78, www.jstor.org/stable/10.2307/27143339.

2. David Western, *We Alone: How Humans Have Conquered the Planet and Can Also Save It* (New Haven, CT: Yale University Press, 2020).

3. Erle C. Ellis, "The Anthropocene Condition: Evolving through Social-Ecological Transformations," *Philosophical Transactions of the Royal Society B: Biological Sciences* 379 (2023): e20220255, https://doi.org/10.1098/rstb.2022.0255.

4. Martin E. P. Seligman, Peter Railton, Roy F. Baumeister, and Chandra Sripada, *Homo Prospectus* (Oxford: Oxford University Press, 2016).

# Index

Footnotes are indicated by "n" following the page number.